Installing Seismic Restraints for Duct and Pipe

FEMA P-414/January 2004

 FEMA

For additional copies of this guide, call the FEMA Distribution Center: 1-800-480-2520.

For further information concerning this guide or the activities of ASCE, contact the American Society of Civil Engineers, 1801 Alexander Bell Drive, Reston, Virginia, 20191, 703-295-6000.

For further information concerning the activities of the Vibration Isolation and Seismic Control Manufacturers Association, contact Mr. Robert Ecker, 610-971-4850.

All photos used with permission. Photos provided by:

    Air Balance
    American Standard, Inc.
    California Economizer, Inc.
    DRI-STREEM Humidifier Company
    Emerson Process Management / Fisher Controls
    Greenheck Fan Corporation
    ITT Industries / Bell and Gossett
    KINETICS Noise Control

Installing Seismic Restraints for Duct and Pipe

FEMA P-414/January 2004

Developed by the Vibration Isolation and Seismic Control Manufacturers Association under a cooperative agreement between the Federal Emergency Management Agency and the American Society of Civil Engineers

ACKNOWLEDGEMENTS

This guide was prepared under Cooperative Agreement EMW-2001-CO-0379 between the Federal Emergency Management Agency and the American Society of Civil Engineers. The contributions of the following are gratefully acknowledged:

Vibration Isolation and Seismic Control Manufacturers Association (VISCMA)

Amber/Booth Company
Kinetics Noise Control
Mason Industries
M. W. Sausse & Company

Thybar Corporation
Vibration Mountings & Controls, Inc.
Vibration Eliminator Co., Inc.

Concrete Anchors Manufacturers Association (CAMA)

Sheet Metal and Air Conditioning Contractors Association, Inc. (SMACNA)

The American Society of Civil Engineers (ASCE)

Project Team

Paul Tertell, FEMA Project Officer
Suzanne Ramsey, Project Manager
James Carlson, Principal Investigator

J. Marx Ayres
James Borchers
Mike Callanan
Robert H. Ecker

Michael Hagerty
James Henderson
Jerry Lilly
Bill Staehlin

Consulting Advisors - Representatives of

Amber/Booth Company
IAPMO
Kinetics Noise Control
Mason Industries, Inc.
M. W. Sausse & Co., Inc.

Thybar Corporation
Vibration Mountings & Controls, Inc.
Vibration Eliminator Co., Inc.
Wej-it Fastening Systems

Advisory Group

Doug Fitts
Jack Ivers
Brett McCord
Gary McGavin

Editing and Layout

Gary Clemmons, GC Ink
Leslie McCasker, McCasker Consulting

Illustrations

Scott Hiner, Creative Marketing Resources

The Federal Emergency Management Agency (FEMA)

TABLE OF CONTENTS

Table of Contents

INTRODUCTION

This guide shows installers how to attach ducts, pipes, and associated equipment to a building to minimize earthquake damage. Many attachment examples and arrangements are presented, including anchors and the use of special devices called *seismic restraint devices*.

Seismic restraint devices include vibration isolation systems, cable or strut suspension systems, roof attachment systems, and the use of steel shapes.

Please note that this guide does not replace:

- Printed instructions shipped with the equipment.
- Instructions in construction documents and specifications. Use approved construction documents.
- Code-required, industry accepted practices.
- Safety guidelines and practices.
- Orders from your supervisor.
- Seismic restraint device submittals.

Please note that this guide does not include fire protection sprinkler, smoke and fire stops, or fire detection governed by local codes and the National Fire Protection Association.

If you have questions about any information in this guide, check with your supervisor.

This guide contains these sections:

- *Bracing Layout and Selection:* Organized by duct and pipe components.
- *Bracing Details and Installation:* Organized by duct and pipe components. Gives instructions on installing bracing in many different arrangements.
- *Attachments:* Contains instructions on attaching suspended equipment and attachment details that typically apply to connecting ducts and pipes to building structures.
- *Anchors:* Shows many different types of anchors used to connect equipment to a building.
- *Special Cases:* Covers cable assemblies, and special situations involving seismic joints, valves and valve actuators.

Introduction

To use this guide:

1. Use the Table of Contents to find the Equipment section that best represents the equipment you are installing.

2. Using the table (see example below) in the Equipment section, find the:
    -   type of equipment you are installing in column 1
    -   method of installing the equipment in column 2
    -   attachment type in column 3.

column 1	column 2	column 3
Typical Equipment	*How is equipment to be installed?*	*Attachment Type*
Any type of unit	Connected to angles mounted to the floor	Rigid with angles *Go to page 53*

3. Turn to the page referenced in column 3 for the equipment/attachment type you have selected. If you are not sure which attachment type is correct, ask your supervisor.

4. Follow the instructions for the attachment type you have selected. These instructions will refer you to the correct anchor section so you can make the connection to the building structure.

NOTE: All instructions in this guide are arranged in order using numbered steps. Please follow every step in the sequence shown.

Special precautions are marked:

 A flag means you should take special care before continuing. Read all the information next to a flag before making the attachment.

 A warning sign means you can cause serious damage to the building, the device, or the equipment if you do not follow the instructions exactly.

 A book means you should refer to the manufacturer's printed instructions before continuing.

Note that a Glossary and an Index are also available to facilitate use of this guide.

DUCT AND PIPE BRACING LAYOUT

 Be sure to refer to approved construction drawings and specifications, seismic restraint submittals, and manufacturer's instructions. Also, refer to the manual that was used to design the seismic bracing. This manual is required to be on the job site.

Refer to approved construction documents that show the overall layout of the duct and pipe runs throughout the building. Normal vertical supports are provided at intervals as defined in codes and standards. Additional seismic bracing may be required.

Hanger and bracing locations are found in approved construction documents. Contact your supervisor to obtain these construction documents. When additional seismic bracing is required, follow the steps in this section to identify where you should place the additional bracing. Then turn to the page showing the details and installation instructions for each type of seismic brace.

Exceptions for ducts and piping: Refer to codes and manufacturer's manuals.

 Refer to adopted local codes for any requirements that must be met to exclude seismic bracing.

 Refer to local codes for hanger spacing requirements.

Duct and Pipe Bracing Layout

A run is a single straight section of duct or piping. Any change in direction is considered a new or different run. Offsets within a run may be allowed if the offset is less than the recommended spacing divided by 16 or as allowed by approved construction documents.

Separate the layout into runs as shown in Figure 1.

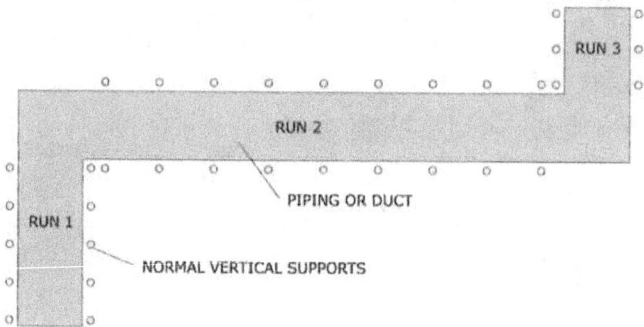

Figure 1: Separate layout into runs.

For bracing a single pipe run, there may be many short sections of pipe. Single or multiple offsets may be allowed if the total offset is less than the recommended transverse spacing divided by 16.

Offsets are shown in Figure 2.

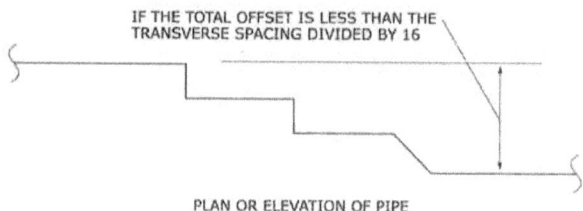

Figure 2: Pipe offset exception.

Step 2: Lay out transverse bracing

At a minimum, transverse bracing must be located at each end of the run as shown in Figure 3.

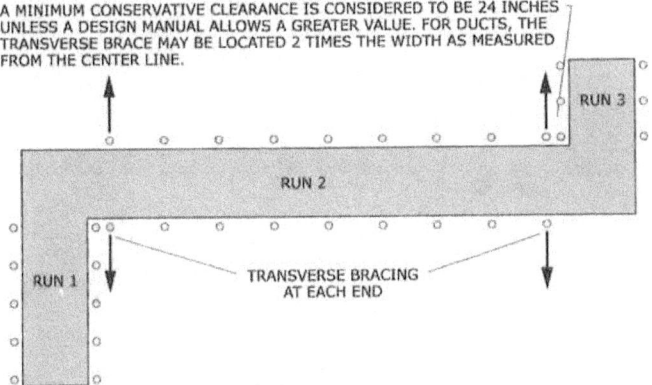

A MINIMUM CONSERVATIVE CLEARANCE IS CONSIDERED TO BE 24 INCHES UNLESS A DESIGN MANUAL ALLOWS A GREATER VALUE. FOR DUCTS, THE TRANSVERSE BRACE MAY BE LOCATED 2 TIMES THE WIDTH AS MEASURED FROM THE CENTER LINE.

Figure 3: Locate transverse bracing at the ends.

Step 3: Check to see if additional transverse bracing is required

 Refer to the bracing manual at the job site.

If the length of the run is greater than the allowed transverse spacing in the bracing manual, add intermediate transverse bracing until the spacing is correct as shown in Figure 4.

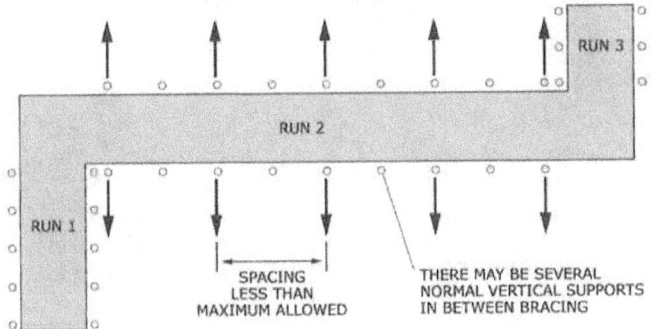

Figure 4: Install additional transverse bracing as necessary, to stay within maximum transverse spacing limitations.

Step 4: Add longitudinal bracing

Each run must have a minimum of one longitudinal brace as shown in Figure 5.

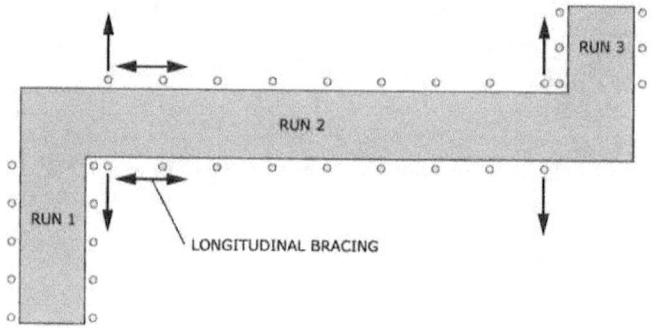

Figure 5: Add longitudinal bracing as necessary.

Transverse bracing on adjacent runs may be considered the longitudinal bracing as shown in Figure 6.

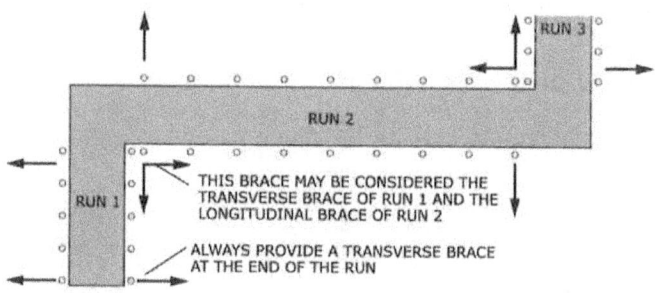

Figure 6: One cost-effective way to rearrange bracing.

Longitudinal bracing is usually spaced at a maximum distance that is two times the transverse spacing. For long runs, every second transverse brace should also have a longitudinal brace, as shown in Figure 7 (page 7).

Sometimes longitudinal bracing may be required on every transverse brace for large ducts and pipes.

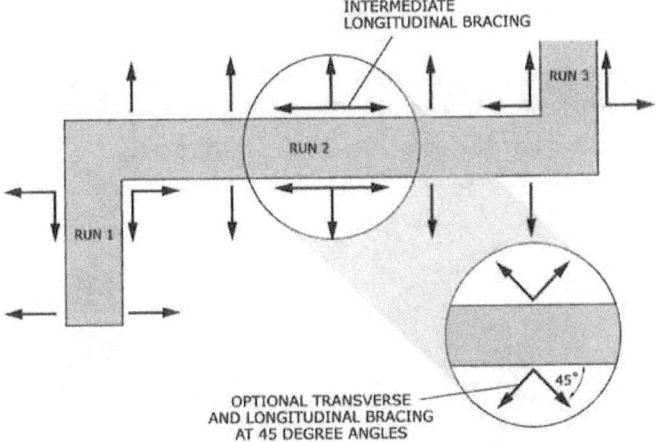

Figure 7: Final configuration.

Refer to the manufacturer's/industry manual for bracing requirements as designated in approved construction documents.

Refer to Figure 67 (page 63) for optional transverse and longitudinal bracing.

All longitudinal-only bracing must be physically attached to pipe. Attaching clevis-type hangers are not acceptable.

Go to pages 8-15 to select the bracing type for typical installations. Bracing details start on page 16.

DUCT/PIPE/IN-LINE EQUIPMENT BRACING SELECTION

Duct Bracing Selection

 Be sure to refer to approved construction drawings and specifications, seismic restraint submittals, and manufacturer's instructions.

Step 1: Identify the duct bracing used

Figure 8 shows the different ways to brace ducts.

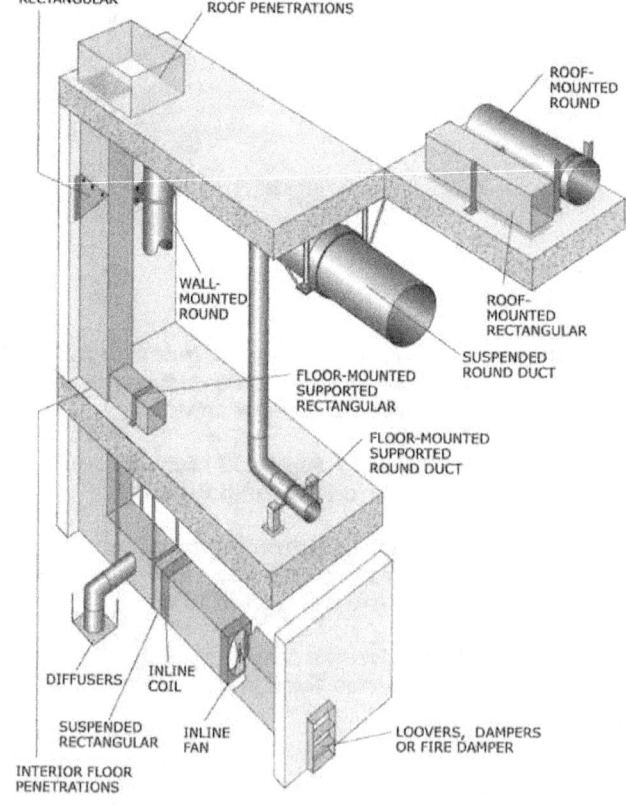

Figure 8: Duct system with different seismic braces.

In-line duct-mounted equipment can be found on page 10.

Step 2: Select the bracing type

Using the following table, select the attachment that best matches the installation you have selected, then turn to the page listed under the bracing type.

Typical Arrangements	How is equipment to be installed?	Bracing Type
Suspended rectangular duct	Suspended with rods or angles using cables or rigid laterals	Suspended rectangular ducts *Go to page 16*
	Isolated with rod supports and cable bracing	Vibration-isolated rectangular ducts *Go to page 26*
Suspended round duct (including oval duct)	Suspended with rods or angles using cables or rigid laterals	Suspended round ducts *Go to page 28*
	Isolated with rod supports and cable bracing	Vibration-isolated round ducts *Go to page 39*
Floor-mounted rectangular and round duct	Suspended off the floor with angles	Floor-mounted ducts *Go to page 41*
Roof-mounted rectangular and round duct	Suspended above the roof with angles	Roof mounted ducts *Go to page 44*
Wall-mounted rectangular and round duct	Supported off the wall	Wall mounted ducts *Go to page 47*
Duct penetrations	Ducts through building structure	Duct Penetrations *Go to page 50*
In-line duct-mounted equipment	Attached to the building structure	In-line duct - mounted equipment *Go to page 10*

Table 1: Duct bracing installation types.

 Refer to approved contract documents for details and provisions for crossing fire barriers, area separation walls/floors/roofs, smoke barriers and seismic separation joints.

 Refer to approved contract documents for re uirements for weatherproofing roof and/or floor penetrations.

In-line Duct-mounted Equipment Bracing Selection

 Be sure to refer to construction drawings and specifications, seismic restraint submittals, and manufacturer's instructions.

Step 1: Identify the in-line duct-mounted equipment

Figure 9: Motorized damper.

Figure 10: Motorized damper.

Figure 11: Combination smoke and fire damper.

Figure 12: Square duct-mounted fire damper.

Figure 13: Round fire damper.

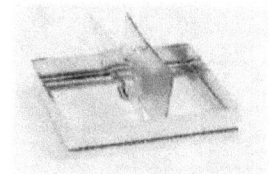

Figure 14: Ceiling fire damper.

Figure 15: Coil.

Figure 16: Humidifier.

Figure 17: Silencer.

Figure 18: Fan-coil unit.

Figure 19: VAV terminal unit.

Figure 20: In-line fan.

Step 2: Select the attachment type

Using the following table, select the attachment that best
matches the installation you have selected, then turn to the
page listed under the attachment type.

Typical Arrangements	How is equipment to be installed?	Attachment Type
Coil, damper, in-line fan, fan-coil unit, duct silencer, or VAV terminal unit	Suspended from the structure above using rods and cables	Rigid _Go to page 85_
	Suspended from the structure above using isolators, rods and cables	Isolated _Go to page 90_
	Suspended from the structure above using angles	Rigid with angles _Go to page 88_
Humidifier	Supported by the duct wall	Follow manufacturer's instructions
	Suspended from the structure	

Table 2: In-line duct-mounted equipment installation types.

Pipe Bracing Selection

 Be sure to refer to approved construction drawings and specifications, seismic restraint submittals, and manufacturer's instructions.

Step 1: Identify the pipe bracing used

Figure 21 shows the different ways to brace pipes.

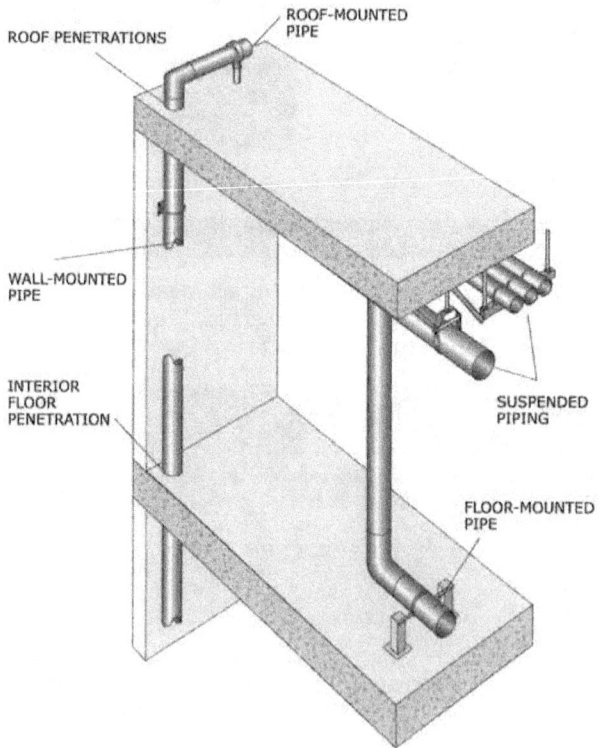

ROOF PENETRATIONS

ROOF-MOUNTED PIPE

WALL-MOUNTED PIPE

INTERIOR FLOOR PENETRATION

SUSPENDED PIPING

FLOOR-MOUNTED PIPE

Figure 21: Pipe system with different seismic bracing.

In-line pipe-mounted equipment can be found on page 14.

Step 2: Select the bracing type

Using the following table, select the attachment that best matches the installation you have selected, then turn to the page listed under the bracing type.

Typical Arrangements	How is equipment to be installed?	Bracing Type
Suspended piping	Suspended with rods and angles using cables or rigid laterals	Suspended piping *Go to page 54*
	Isolated with rod supports and cable bracing	Vibration-isolated *Go to page 65*
Floor-mounted piping	Suspended off the floor with angles/struts	Floor-mounted piping *Go to page 70*
Roof-mounted piping	Braced above the roof	Roof mounted piping *Go to page 73*
Wall-mounted piping	Supported off the wall	Wall-mounted piping *Go to page 76*
Pipe penetrations	Pipes through building structure	Pipe penetrations *Go to page 79*
In-line pipe-mounted equipment	Attached to the building structure	In-line pipe - mounted equipment *Go to page 14*

Table 3: Pipe bracing installation types.

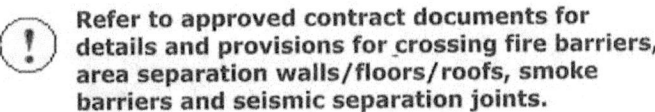

Refer to approved contract documents for details and provisions for crossing fire barriers, area separation walls/floors/roofs, smoke barriers and seismic separation joints.

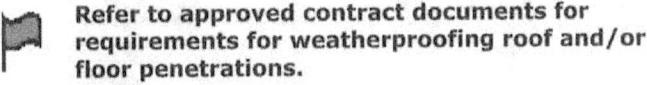

Refer to approved contract documents for requirements for weatherproofing roof and/or floor penetrations.

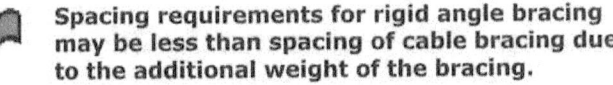

Spacing requirements for rigid angle bracing may be less than spacing of cable bracing due to the additional weight of the bracing.

In-line Pipe-mounted Equipment Bracing Selection

 Be sure to refer to approved construction drawings and specifications, seismic restraint submittals, and manufacturer's instructions.

Step 1: Identify the in-line pipe-mounted equipment

Figure 22: Valves and valve air actuator.

Figure 23: Valve and electronic actuator.

Figure 24: In-line pump.

Figure 25: Air separator.

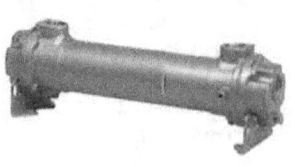

Figure 26: Heat exchanger.

Figure 27: Strainer.

Step 2: Select the attachment type

Using the following table, select the attachment that best matches the installation you have selected, then turn to the page listed under the attachment type.

Typical Arrangements	How is equipment to be installed?	Attachment Type
Valve or strainer	Requires additional bracing at the valve and actuator if it weighs more than 20 pounds	Valves and valve actuators *Go to page 144*
Valve actuator	Requires additional bracing if it weighs more than 20 pounds	
Air separator, in-line pump or heat exchanger	Support piping near equipment/valve	Suspended piping and pumps *Go to page 97* Air separator *Go to page 99* Heat exchanger *Go to page 100*

Table 4: In-line pipe-mounted equipment installation types.

 Valves with brittle valve bodies/connections require bracing near the valve.

 Strainers and other piping specialties are braced similar to valves.

DUCT BRACING DETAILS AND INSTALLATION INSTRUCTIONS

This section gives instructions on bracing six different kinds of ducts:

- Suspended rectangular ducts (this page).
- Suspended round ducts (page 28).
- Floor-mounted ducts (page 41).
- Roof-mounted ducts (page 44).
- Wall- and chase-mounted ducts (page 47).
- Duct penetrations (page 50).

Suspended Rectangular Ducts

The six ways to brace suspended rectangular ducts are by using:

- Vertical rods with cable bracing (page 17).
- Vertical rods with steel-shaped bracing (page 20).
- Vertical steel shapes with cable bracing (page 22).
- Vertical steel shapes with steel-shaped bracing (page 23).
- Unbraced supports (page 24).
- Vibration-isolated rectangular duct (page 26).

For post-tension (pre-stressed) buildings, locate the tendons before drilling. Extreme damage may occur if a tendon is nicked or cut.

Refer to approved construction documents for details and provisions for crossing fire barriers, area separation walls/floors/roofs, smoke barriers, and seismic separation joints.

Pre-approved manufacturer's/industry manuals used for the installation of duct and pipe bracing are required to be on the job site to ensure that the correct details are being used.

Bracing must not be attaced to duct joint.

16

Vertical rods with cable bracing

Vertical rod-braced ducts with transverse and longitudinal supports are shown in Figure 28 (below), Figure 29 (page 18), and Figure 30 (page 19).

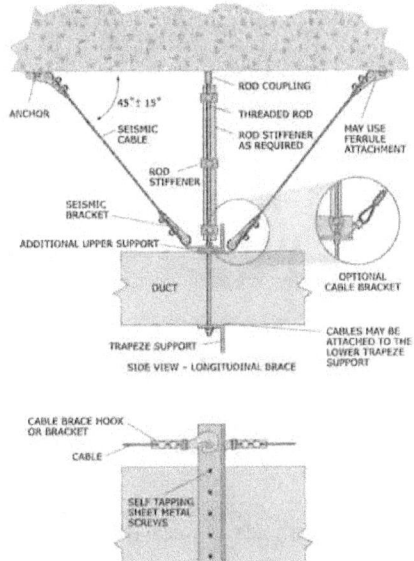

Figure 28: Rectangular duct with vertical rods and braced with cables (longitudinal).

> 🚩 **Use the shortest screws possible when penetrating ductwork to minimize airflow noise inside the duct.**

> 🚩 **Do not use coupling nuts except at building structure connection.**

> 🚩 **Longitudinal cable bracing requires bracing in both directions (as shown).**

> 🚩 **Rod must be one piece from building structure to bottom of trapeze support.**

Bracing Details and Installation Instructions: Suspended Rectangular Ducts

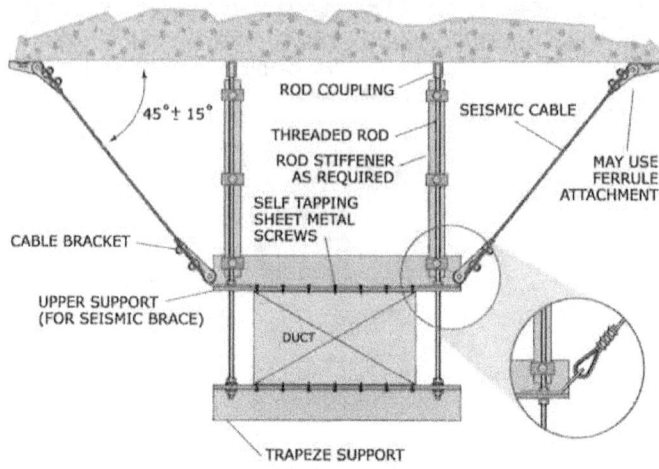

FRONT VIEW - TRANSVERSE BRACE

Figure 29: Rectangular duct with vertical rods braced with cables (transverse).

 Rod must be one piece from building structure to bottom of trapeze support.

Step 1: Attach vertical rods with hanger to the building structure

Lay out all attachment points before anchoring, then refer to Attachment Details Connecting to Building Structure (page 102). For instructions on installing anchors, see Anchors (page 107).

 The building structure must be point-load capable. Verify with the appropriate design professional.

Step 2: Run duct as required by approved construction documents

Assemble the support and connect it to vertical rods as shown in Figure 28 (page 17), Figure 29 (above), and Figure 30 (page 19).

18

Step 3: Install anchors for bracing

Install sheet metal screws to secure duct. Install brackets and cable bracing. For cable assembly instructions, see Cables (page 134). For instructions on installing anchors, see Anchors (page 107).

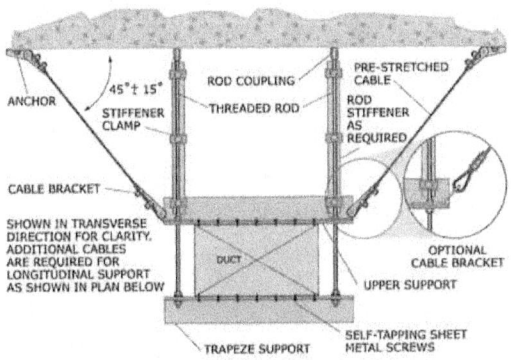

FRONT VIEW – ALL DIRECTIONAL BRACE

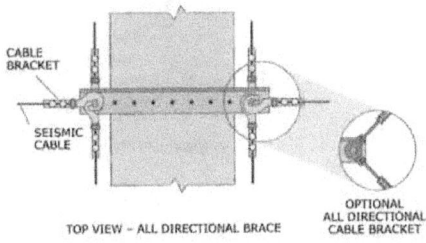

TOP VIEW – ALL DIRECTIONAL BRACE

Figure 30: Rectangular duct with vertical rods and braced with cables (all directional).

 Rod must be one piece from building structure to bottom of trapeze support.

 Do not attach threaded rod and cable to the same anchor.

END OF DETAIL.

Vertical rods with steel shaped bracing

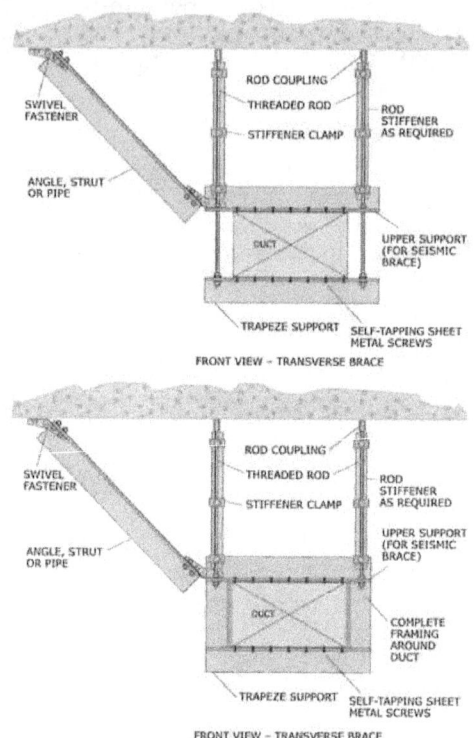

Figure 31: Rectangular duct with vertical rods and braced with steel shapes (transverse).

 Rod must be one piece from building structure to bottom of trapeze support.

 Pre-approved manufacturer's/industry manuals may limit the maximum transverse and longitudinal angles to 45 degrees.

 Additonal weight of steel shaped bracing added to dead load and seismic lateral load may reduce brace spacing or increase rod and rod anchor sizes.

Step 1: Attach vertical rods with hanger to the building structure

Lay out all attachment points before anchoring, then refer to Attachment Details Connecting to Building Structure (page 102). For instructions on installing anchors, see Anchors (page 107).

 Building structure must be point-load capable. Verify with the appropriate design professional.

Step 2: Run duct as required by approved construction documents

Assemble the support and connect to vertical rods as shown in the Figure 31 (page 20).

Step 3: Install anchors for bracing

Install brackets and cable bracing to building structure and support. For cable assembly instructions, see Cables (page 134). For instructions on installing anchors, see Anchors (page 107).

END OF DETAIL.

Bracing Details and Installation Instructions: Suspended Rectangular Ducts

Vertical steel shapes with cable bracing

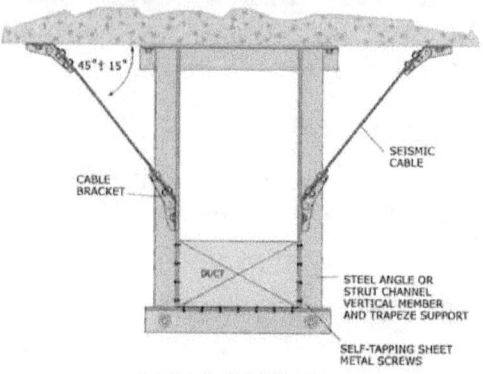

FRONT VIEW – TRANSVERSE BRACE

Figure 32: Rectangular duct with steel shapes and braced with cables (transverse).

Step 1: Attach vertical steel shapes to the building structure

Lay out all attachment points before anchoring, then refer to Attachment Details Connecting to Building Structure (page 102). For instructions on installing anchors, see Anchors (page 107).

 Building structure must be point-load capable. Verify with the appropriate design professional.

Step 2: Run duct as required by approved construction documents

Assemble the support and connect to vertical rods as shown in the Figure 32 (above).

Step 3: Install anchors for bracing

Install brackets and cable bracing to building structure and support. For cable assembly instructions, see Cables (page 134). For instructions on installing anchors, see Anchors (page 107).

END OF DETAIL.

Vertical steel shapes with steel shaped bracing

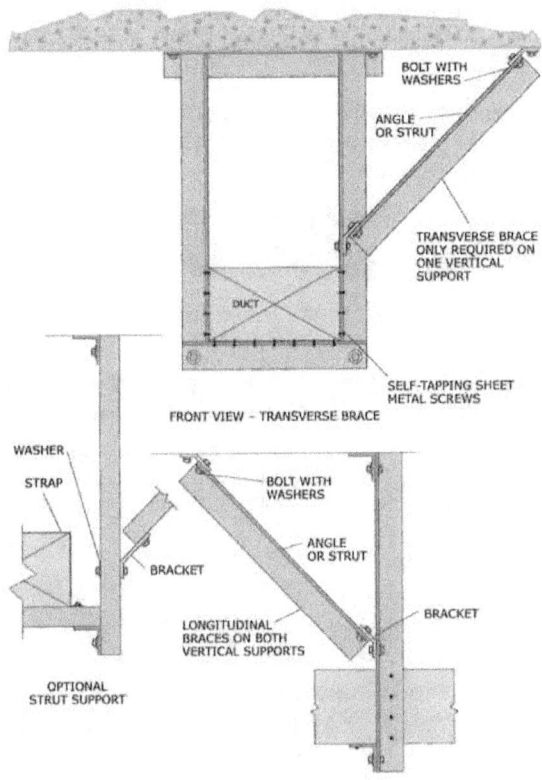

Figure 33: Rectangular duct with vertical steel shapes and steel shaped bracing (transverse and longitudinal).

 Rod must be one piece from building structure to bottom of trapeze support.

 Pre-approved manufacturer's/industry manuals may limit the maximum transverse and longitudinal angles to 45 degrees.

Bracing Details and Installation Instructions: Suspended
Rectangular Ducts

Step 1: Attach vertical rods to the building structure

Lay out all attachment points before anchoring, then refer to
Attachment Details Connecting to Building Structure (page
102).

 **Building structure must be point-load capable.
Verify with the appropriate design
professional.**

Step 2: Run duct as required by approved construction documents

Assemble duct bracing and connect to vertical angles as
shown in the Figure 33 (page 23).

Step 3: Install anchors for bracing

Install brackets and cable bracing to building structure and
support. For cable assembly instructions, see Cables (page
134). For instructions on installing anchors, see Anchors
(page 107).

END OF DETAIL.

Unbraced supports

 **Unbraced supports may be allowed by the
authority having jurisdiction and may require
the top of the duct to be attached to the
support.**

 **Refer to building codes for required exceptions
to unbraced piping and ducts.**

There are three types of unbraced supports:

- Non-moment-resistant rod support
- Strap support
- Vibration-isolated support (see the optional view in
  Figure 34 on page 25)

Non-moment-resistant rod support

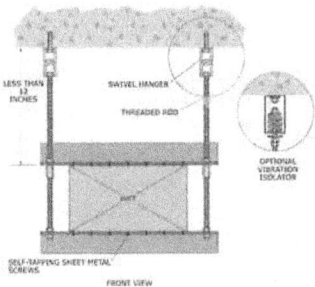

Figure 34: Rectangular duct non-moment-resistant rod support.

Strap support

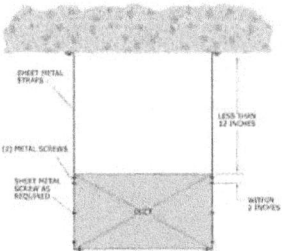

Figure 35: Rectangular duct strap support.

 Refer to manufacturer's/industry manuals for size of hanger supports (straps) and spacing.

Step 1: Attach rods/straps to the building structure with anchors

Lay out all attachment points before anchoring. Attach as shown in Figure 34 (above) and Figure 35 (above).

 Building structure must be point-load capable. Verify with the appropriate design professional.

Step 2: Run duct as required by approved construction documents

Connect straps to duct as shown in Figure 35 (above) or assemble duct rod support as shown in Figure 36 (page 26).

END OF DETAIL.

Bracing Details and Installation Instructions: Suspended Rectangular Ducts

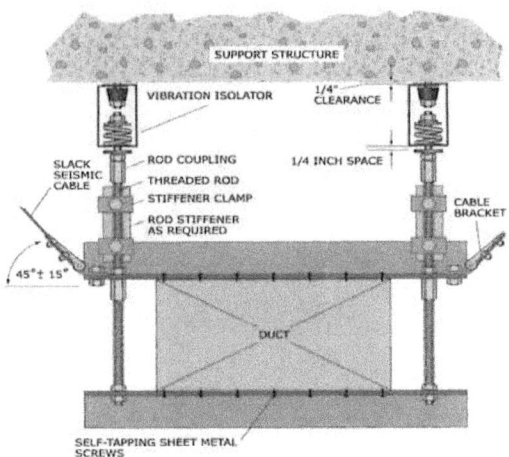

FRONT VIEW – TRANSVERSE BRACE

Figure 36: Rectangular duct isolated with vertical rods and braced with cables (transverse).

 Verify that the vertical limit stops and clearances meet the manufacturer's requirements.

Step 1: Attach vertical rods with vibration isolators to the building structure

Lay out all the attachment points before anchoring, then refer to Attachment Details Connecting to Building Structure (page 102). For isolator details, refer to Figure 92 (page 91). For instructions on installing anchors, see Anchors (page 107).

 Building structure must be point-load capable. Verify with the appropriate design professional.

Step 2: Run duct as required by approved construction documents

Assemble angle bracing and connect to vertical rods as shown in Figure 36 (above).

Step 3: Install anchors for bracing

Install brackets and cable bracing to building structure and support. For cable assembly instructions, see Cables (page 134). For instructions on installing anchors, see Anchors (page 107).

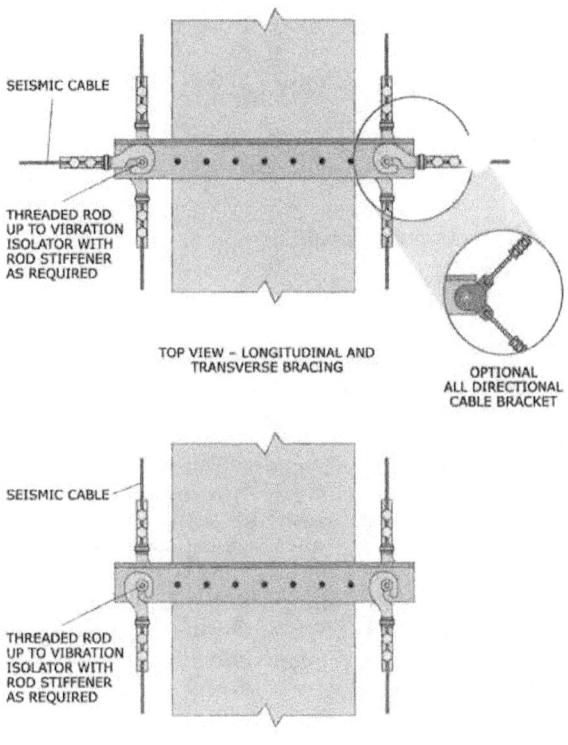

Figure 37: Rectangular duct isolated with rods and braced with cables (all directional and longitudinal).

END OF DETAIL.

Suspended Round Ducts

The six ways to brace suspended round and oval duct supports are by using:

- Vertical rods with cable bracing (page 29).

- Vertical rods with steel shaped bracing (page 32).

- Vertical steel shapes with cable bracing (page 34).

- Vertical steel shapes with steel shaped bracing (page 36).

- Unbraced supports (page 37).

- Vibration isolation (page 39).

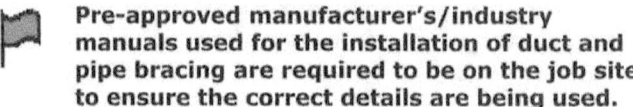 **For post-tension (pre-stressed) buildings, locate the tendons before drilling. Extreme damage may occur if a tendon is nicked or cut.**

Refer to approved construction documents for details and provisions for crossing fire barriers, area separation walls/floors/roofs, smoke barriers, and seismic separation joints.

Pre-approved manufacturer's/industry manuals used for the installation of duct and pipe bracing are required to be on the job site to ensure the correct details are being used.

Vertical rods with cable bracing

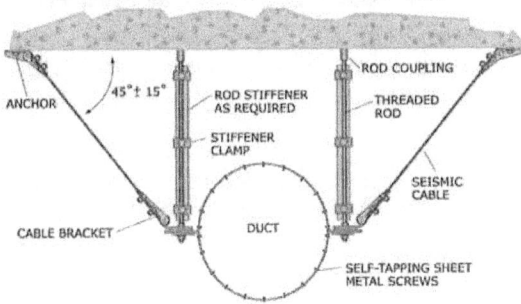

FRONT VIEW – TRANSVERSE BRACE

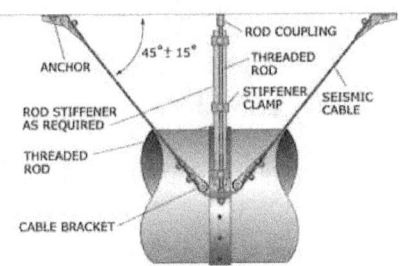

SIDE VIEW – LONGITUDINAL BRACE

Figure 38: Round duct with vertical rods and braced with cables
(transverse and longitudinal).

 **Use the shortest screws possible when
penetrating ductwork to minimize airflow noise
inside the duct.**

 **Cable bracing requires bracing in both
directions as shown in Figure 38 (above).**

Step 1: Attach vertical rods with hanger to the building structure

Lay out all attachment points before anchoring, then refer to
Attachment Details Connecting to Building Structure (page
102). For instructions on installing anchors, see Anchors
(page 107).

 **Building structure must be point-load capable.
Verify with the appropriate design professional.**

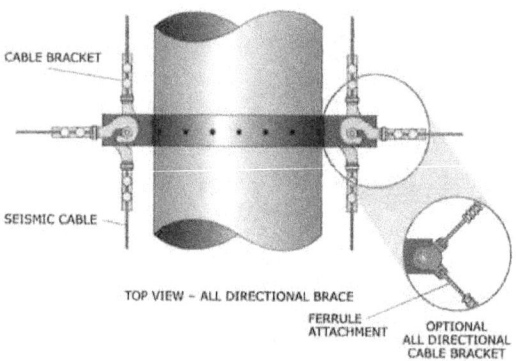

TOP VIEW – ALL DIRECTIONAL BRACE

Figure 39: Round duct with vertical rods and braced with cables (all
directional).

 **Use the shortest screws possible when
penetrating ductwork to minimize airflow noise
inside the duct.**

 **Cable bracing requires bracing on both sides as
shown in Figure 39 (above) and Figure 40
(page 31).**

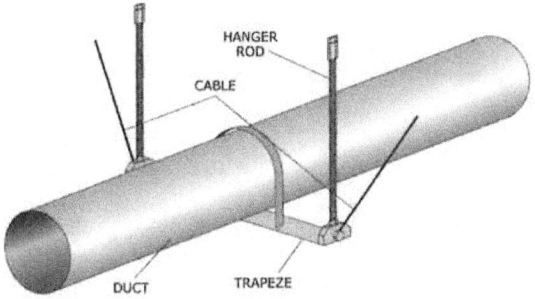

Figure 40: Optional round duct with vertical rods and braced with cables (transverse).

Step 2: Run duct as required by approved construction documents

Assemble the support and connect it to the vertical rods as shown in Figure 39 (page 30) and Figure 40 (above).

Step 3: Install anchors for bracing

Install brackets and cable bracing to building structure and support. For cable assembly instructions, see Cables (page 134). For instructions on installing anchors, see Anchors (page 107).

END OF DETAIL.

Vertical rods with steel shaped bracing

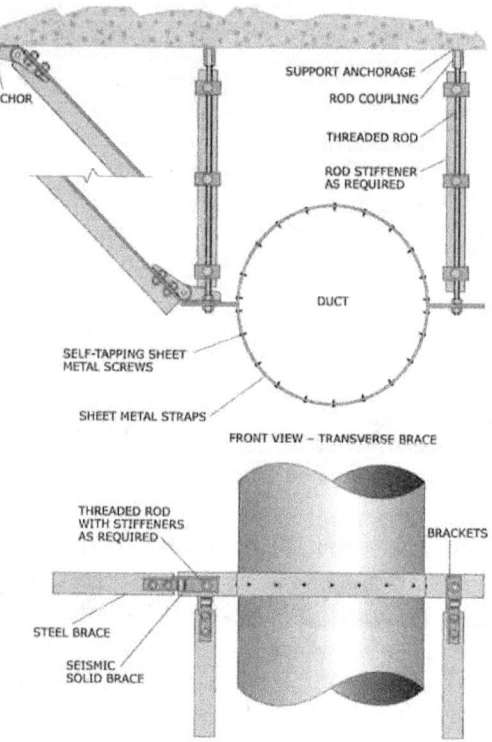

FRONT VIEW – TRANSVERSE BRACE

TOP VIEW – TRANSVERSE AND LONGITUDINAL BRACES

Figure 41: Round duct with vertical rods and braced with steel shapes (transverse and all-directional).

 Pre-approved manufacturer's/industry manuals may limit the maximum transverse and longitudinal angles to 45 degrees.

Step 1: Attach vertical rods with hanger to the building structure

Lay out all attachment points before anchoring, then refer to Attachment Details Connecting to Building Structure (page 102). For instructions on installing anchors, see Anchors (page 107).

 Building structure must be point-load capable. Verify with the appropriate design professional.

Step 2: Run duct as required by approved construction documents

Assemble angle bracing and connect to vertical rods as shown in Figure 41 (page 32).

Step 3: Install anchors for bracing

Install brackets and cable bracing to building structure and support. For cable assembly instructions, see Cables (page 134). For instructions on installing anchors, see Anchors (page 107).

END OF DETAIL.

Vertical steel shapes with cable bracing

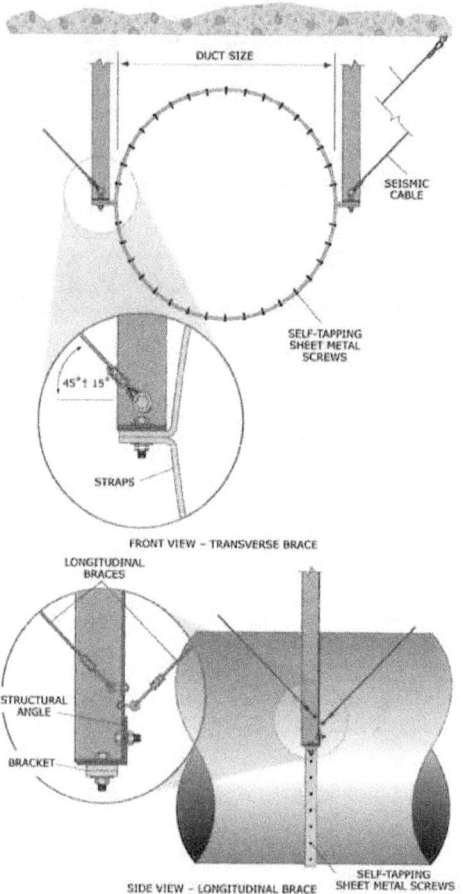

Figure 42: Round duct with vertical steel shapes and braced with
cables (transverse and longitudinal).

 **Use the shortest screws possible when
penetrating ductwork to minimize airflow
noise inside the duct.**

 **Cable bracing requires bracing on both sides as
shown in Figure 42 (above).**

Step 1: Attach vertical steel shapes with hanger to the building structure

Lay out all attachment points before anchoring, then refer to Attachment Details Connecting to Building Structure (page 102). For instructions on installing anchors, see Anchors (page 107).

 Building structure must be point-load capable. Verify with the appropriate design professional.

Step 2: Run duct as required by approved construction documents

Assemble the support and connect it to vertical steel shapes as shown in Figure 42 (page 34).

Step 3: Install anchors for bracing

Install brackets and cable bracing to building structure and support. For cable assembly instructions, see Cables (page 134). For instructions on installing anchors, see Anchors (page 107).

END OF DETAIL.

Vertical steel shapes with steel shaped bracing

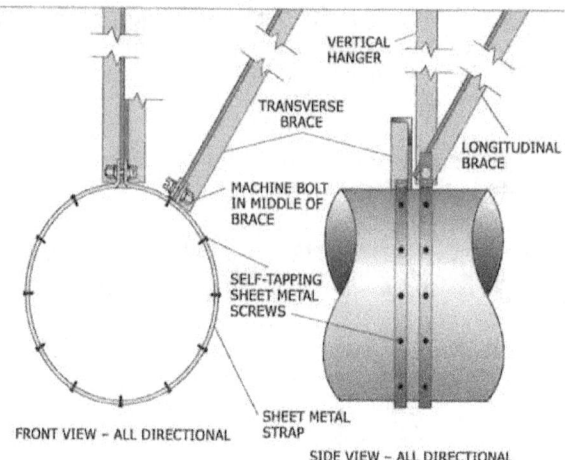

FRONT VIEW – ALL DIRECTIONAL

SIDE VIEW – ALL DIRECTIONAL

Figure 43: Round duct with vertical steel shapes with steel shaped bracing (all-directional).

 Pre-approved manufacturer's/industry manuals may limit the maximum transverse and longitudinal angles to 45 degrees.

Step 1: Attach vertical steel shapes to the building structure

Lay out all attachment points before anchoring, then refer to Attachment Details Connecting to Building Structure (page 102).

 Building structure must be point-load capable. Verify with the appropriate design professional.

Step 2: Run duct as required by approved construction documents

Assemble duct bracing and connect to vertical angles as shown in Figure 43 (above).

Step 3: Install anchors for bracing

Install brackets and cable bracing to building structure and support. For cable assembly instructions, see Cables (page 134). For instructions on installing anchors, see Anchors (page 107).

END OF DETAIL.

Unbraced supports

 Unbraced supports may be allowed by the authority having jurisdiction and may require the top of the duct to be attached to the support.

 Refer to building codes for required exceptions to unbraced piping and ducts.

There are three types of unbraced supports:

* Non-moment-resistant rod support
* Strap support
* Vibration-isolated support (see the optional view in Figure 44 below)

Non-moment-resistant rod support

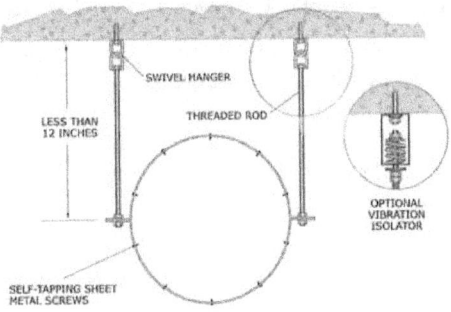

SWIVEL HANGER

LESS THAN
12 INCHES

THREADED ROD

OPTIONAL
VIBRATION
ISOLATOR

SELF-TAPPING SHEET
METAL SCREWS

Figure 44: Round duct non-moment-resistant rod support.

 Use the shortest screws possible when penetrating ductwork to minimize airflow noise inside the duct.

37

Bracing Details and Installation Instructions: Suspended
Round Ducts

Strap Support

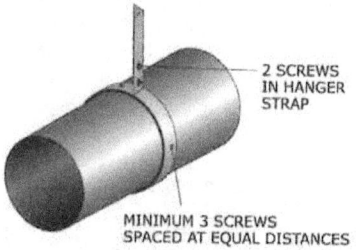

2 SCREWS
IN HANGER
STRAP

MINIMUM 3 SCREWS
SPACED AT EQUAL DISTANCES

Figure 45: Strap support for round duct with steel shaped bracing.

 Refer to manufacturer's/industry manuals for size of hanger supports (straps) and spacing.

Step 1: Attach rods/straps to the building structure with anchors

Lay out all attachment points before anchoring. Attach as shown in Figure 44 (page 37) and Figure 45 (above).

 Building structure must be point-load capable. Verify with the appropriate design professional.

Step 2: Run duct as required by approved construction documents

Connect straps to duct as shown in Figure 45 (above) or assemble duct rod support as shown in Figure 46 (page 39).

END OF DETAIL.

Vibration-isolated round duct

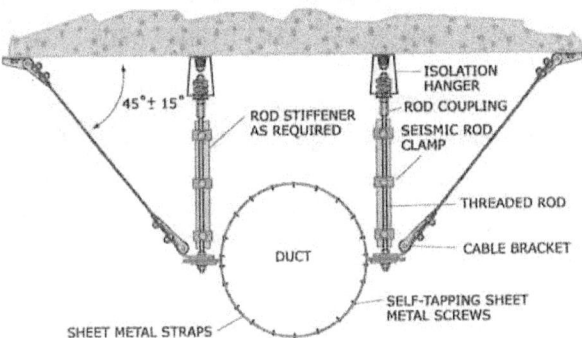

FRONT VIEW – TRANSVERSE BRACING

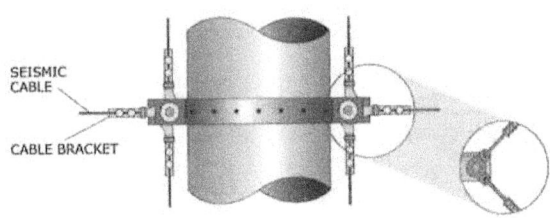

TOP VIEW – ALL DIRECTIONAL BRACING

Figure 46: Round duct isolated with vertical rods and braced with cables (all-directional and transverse).

 Verify that the verical limit stops and clearances meet the manufacturer's requirements.

Step 1: Attach vertical rods with vibration isolators to the building structure

Lay out all attachment points before anchoring, then refer to Attachment Details Connecting to Building Structure (page 102). For instructions on installing anchors, see Anchors (page 107).

 Building structure must be point load capable. Verify with the appropriate design professional.

Bracing Details and Installation Instructions: Suspended
Round Ducts

Step 2: Run duct as required by approved construction documents

Assemble angle bracing and connect to vertical rods as shown in Figure 46 (page 39).

Step 3: Install anchors for bracing

Install brackets and cable bracing to building structure and support. For cable assembly instructions, see Cables (page 134). For instructions on installing anchors, see Anchors (page 107).

END OF DETAIL.

Floor-mounted Ducts

Ducts are usually raised off the floor with a steel shaped
support system. Figure 47 (page 42) and Figure 48 (page
43) show the support with structural steel shapes.

 **For post-tension (pre-stressed) buildings,
locate the tendons before drilling. Extreme
damage may occur if a tendon is nicked or cut.**

 **Refer to approved construction documents for
details and provisions for crossing fire
barriers, area separation walls/floors/roofs,
smoke barriers, and seismic separation joints.**

Step 1: Lay out the duct run

Lay out all attachment points before anchoring, then refer
to Attachment Details Connecting to Building Structure
(page 102).

Step 2: Install anchors for bracing

For instructions on installing anchors, see Anchors (page
107).

Step 3: Assemble angles or straps and secure to anchors and duct with sheet metal screws

Step 4: Run duct as required by approved construction documents

Attach duct to angle assembly with angles and sheet metal
screws. For round ducts, attach duct to assembly with
straps and sheet metal screws.

Bracing Details and Installation Instructions:
Floor-mounted Ducts

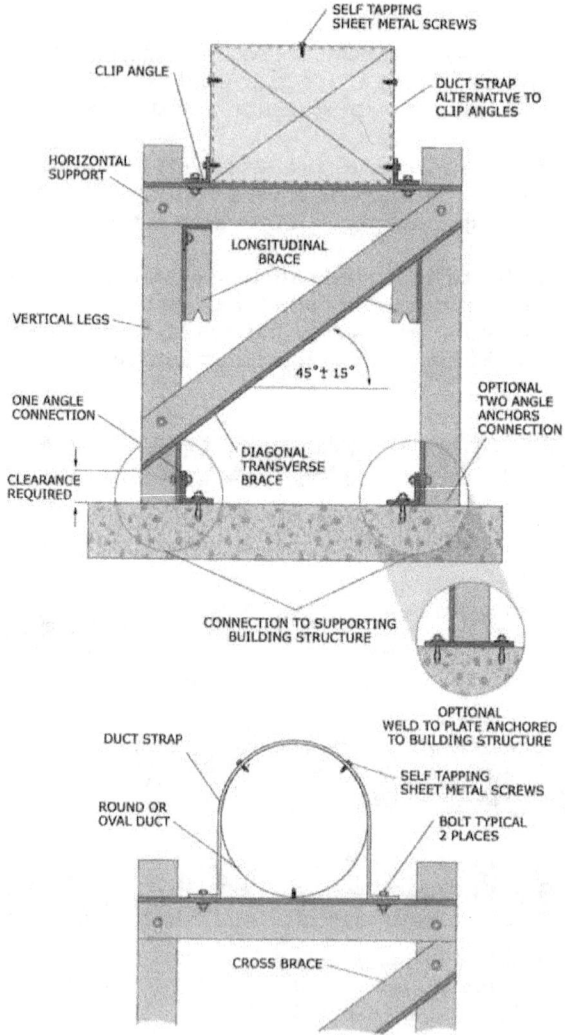

Figure 47: Duct supported off the floor with angles.

 Use the shortest screws possible when penetrating ductwork to minimize airflow noise inside the duct.

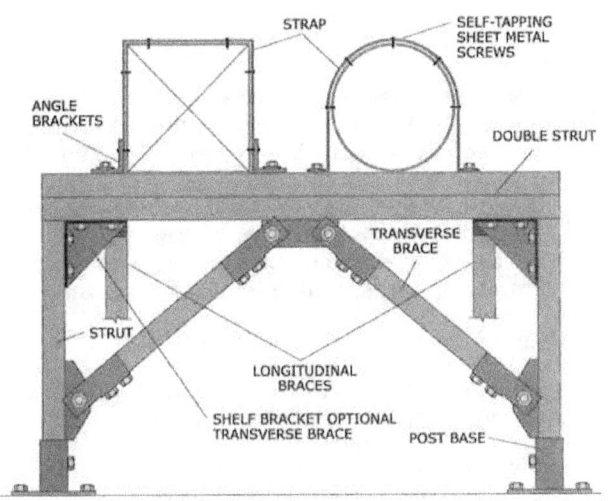

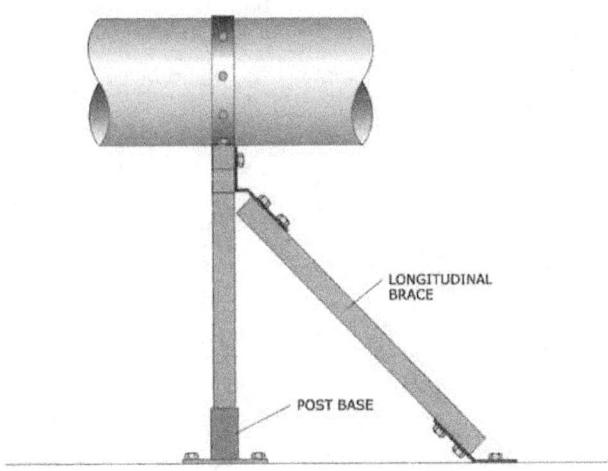

Figure 48: Duct supported off the floor with struts.

 Use the shortest screws possible when penetrating ductwork to minimize airflow noise inside the duct.

Roof-mounted Ducts

Ducts are usually supported above the roof with angles.

> ⚠ **For post-tension (pre-stressed) buildings, locate the tendons before drilling. Extreme damage may occur if a tendon is nicked or cut.**

> ⚠ **Refer to approved construction documents for details and provisions for crossing fire barriers, area separation walls/floors/roofs, smoke barriers, and seismic separation joints.**

The four ways to install roof-mounted ducts are:

- Mounted to a pre-manufactured seismic duct brace (Figure 49, below).

- Mounted to an angle support with cross bracing attached to a roof curb (Figure 50, page 45).

- Mounted to an angle support attached directly to the roof in a pitch pocket to seal the roof (Figure 51, page 45).

- Mounted to an angle support for round or oval ducts (Figure 52, page 46).

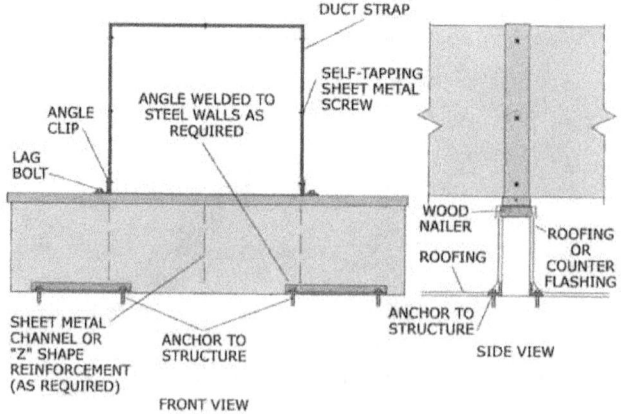

Figure 49: Duct mounted to a pre-manufactured seismic duct brace.

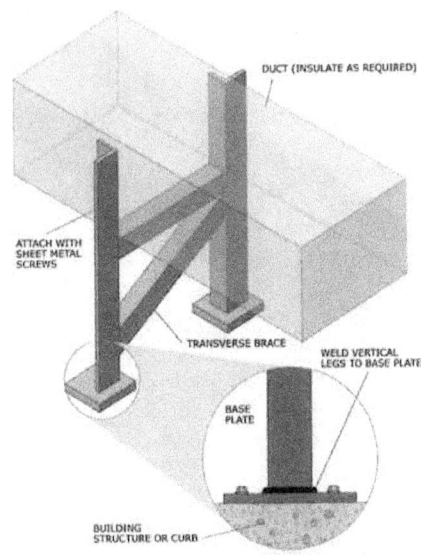

Figure 50: Duct mounted to the roof with cross bracing on a curb.

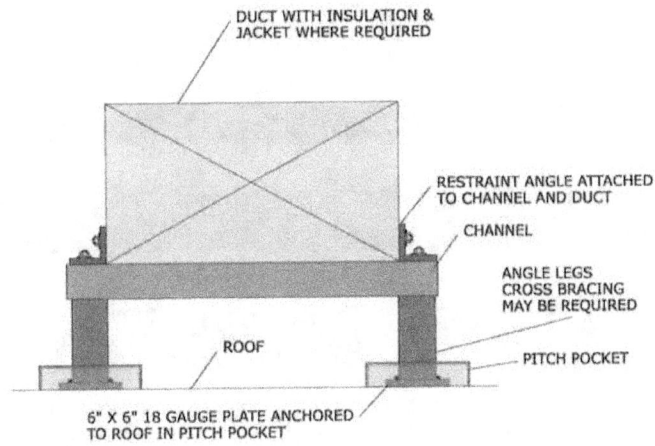

Figure 51: Duct mounted and directly attached to the roof in a pitch pocket.

 Insulate duct with weatherproof jacket where required.

**Bracing Details and Installation Instructions:
Roof-mounted Ducts**

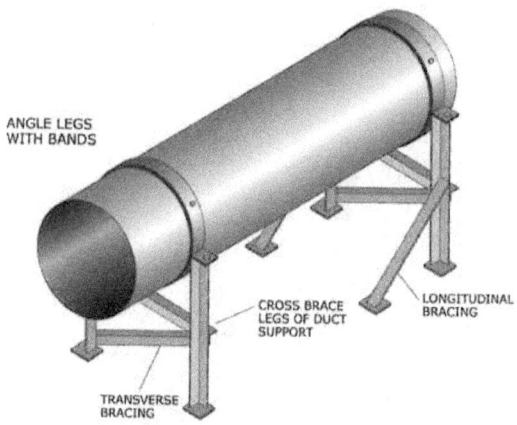

ANGLE LEGS
WITH BANDS

CROSS BRACE
LEGS OF DUCT
SUPPORT

LONGITUDINAL
BRACING

TRANSVERSE
BRACING

Figure 52: Round duct roof support.

Step 1: Lay out the duct run

Lay out all attachment points before anchoring, then refer to
Attachment Details Connecting to Building Structure (page
102). For instructions on installing anchors, see Anchors
(page 107).

Step 2: Assemble support

Step 3: Anchor support base plate to building structure

Step 4: Run duct as required by approved construction documents

Attach duct to the angle assembly with angles and sheet
metal screws. For round ducts, attach duct to the assembly
with straps and sheet metal screws and bolts as required.

 **Seal all anchors and sheet metal screws and
make weatherproof.**

END OF DETAIL.

Wall- and Chase-mounted Ducts

Ducts are usually directly attached to the wall with straps or angles. Ducts can also be supported in a chase.

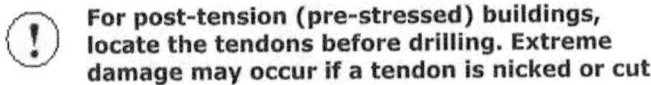 **For post-tension (pre-stressed) buildings, locate the tendons before drilling. Extreme damage may occur if a tendon is nicked or cut.**

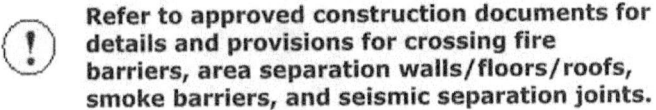 **Refer to approved construction documents for details and provisions for crossing fire barriers, area separation walls/floors/roofs, smoke barriers, and seismic separation joints.**

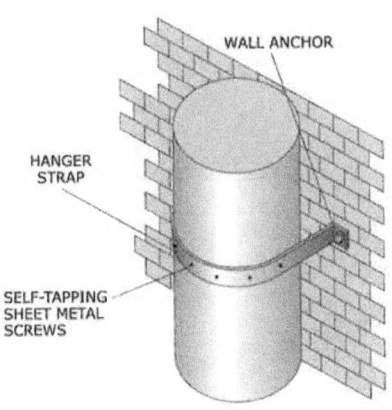

Figure 53: Strap connected directly to wall.

 Straps in Figure 53 (above) do not provide vertical bracing.

 Use the shortest screws possible when penetrating ductwork to minimize airflow noise inside the duct.

**Bracing Details and Installation Instructions:
Wall- and Chase-mounted Ducts**

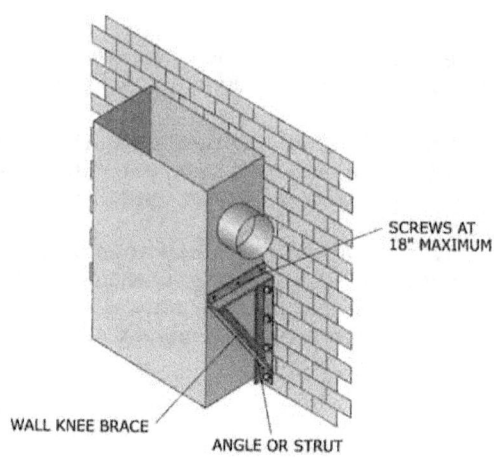

SCREWS AT
18" MAXIMUM

WALL KNEE BRACE

ANGLE OR STRUT

Figure 54: Duct supported by angles.

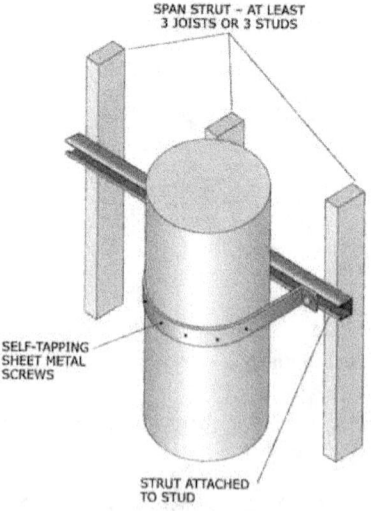

SPAN STRUT – AT LEAST
3 JOISTS OR 3 STUDS

SELF-TAPPING
SHEET METAL
SCREWS

STRUT ATTACHED
TO STUD

Figure 55: Duct supported from wood or metal stud wall.

Bracing Details and Installation Instructions:
Wall- and Chase-mounted Ducts

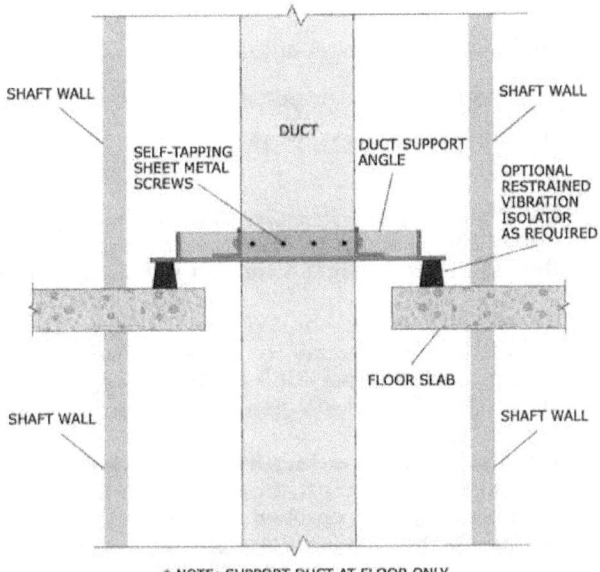

SHAFT WALL

SHAFT WALL

DUCT

SELF-TAPPING
SHEET METAL
SCREWS

DUCT SUPPORT
ANGLE

OPTIONAL
RESTRAINED
VIBRATION
ISOLATOR
AS REQUIRED

FLOOR SLAB

SHAFT WALL

SHAFT WALL

* NOTE: SUPPORT DUCT AT FLOOR ONLY.
DO NOT ATTACH DUCT TO SHAFT WALL.

Figure 56: Duct supported in chase.

Step 1: Lay out the duct run

Lay out all the attachment points before anchoring.

Step 2: Run duct as required by approved construction documents

Step 3: Install anchors for bracing

For instructions on installing anchors, see Anchors (page 107).

Step 4: Assemble angles or straps and secure to anchors and duct with sheet metal screws

 Use the shortest screws possible when penetrating ductwork to minimize airflow noise inside the duct.

END OF DETAIL.

Duct Penetrations

The two types of duct penetrations are:

- Roof duct penetrations (this page).
- Interior duct penetrations (page 52).

> ⚠ **For post-tension (pre-stressed) buildings, locate the tendons before drilling. Extreme damage may occur if a tendon is nicked or cut.**

> ⚠ **Refer to approved construction documents for details and provisions for crossing fire barriers, area separation walls/floors/roofs, smoke barriers, and seismic separation joints.**

> ⚑ **Pre-approved manufacturer's/industry manuals used for the installation of duct and pipe bracing are required to be on the job site to ensure the correct details are being used.**

> ⚑ **All roof penetrations should be sealed and may require flashing.**

Roof duct penetrations

> ⚠ **Coordinate roof penetrations with the roofing contractor.**

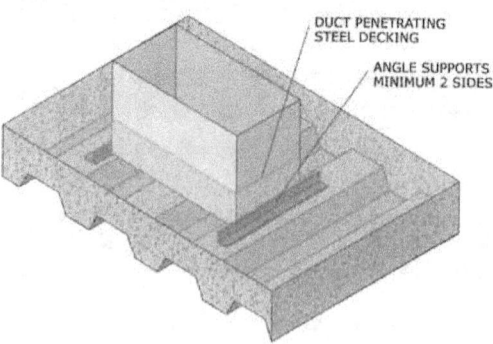

DUCT PENETRATING STEEL DECKING

ANGLE SUPPORTS MINIMUM 2 SIDES

Figure 57: Duct penetration through metal deck roof.

Step 1: Lay out the location of penetration

 Coordinate the layout with a structural engineer. Additional structural supports may be required.

Lay out all attachment points before anchoring.

Step 2: Install anchors for bracing

For instructions on installing anchors, see Anchors (page 107).

Step 3: Run duct as required by approved construction documents

Attach duct to the angle assembly with angles and sheet metal screws. For round ducts, attach duct to the assembly with split band and sheet metal screws and bolts as required. Attach angles to the building structure.

 Use the shortest screws possible when penetrating ductwork to minimize airflow noise inside the duct.

Step 4: Add flashing

 Add flashing as required by approved contract documents or as directed in manufacturer's instructions. See Figure 58 (below).

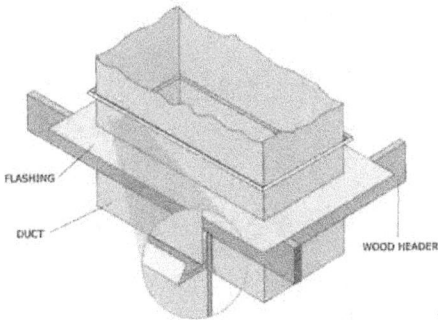

Figure 58: Flashing for roof penetrations.

END OF DETAIL.

Interior duct penetrations

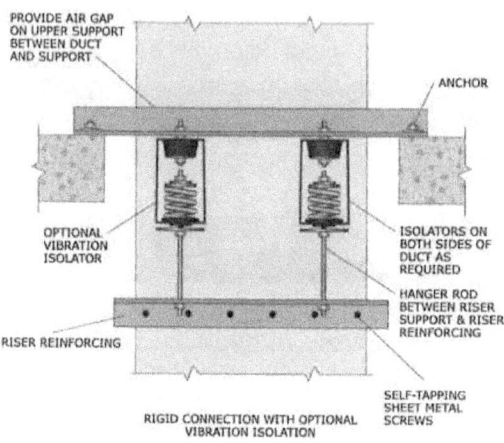

PROVIDE AIR GAP
ON UPPER SUPPORT
BETWEEN DUCT
AND SUPPORT

ANCHOR

OPTIONAL
VIBRATION
ISOLATOR

ISOLATORS ON
BOTH SIDES OF
DUCT AS
REQUIRED

HANGER ROD
BETWEEN RISER
SUPPORT & RISER
REINFORCING

RISER REINFORCING

SELF-TAPPING
SHEET METAL
SCREWS

RIGID CONNECTION WITH OPTIONAL
VIBRATION ISOLATION

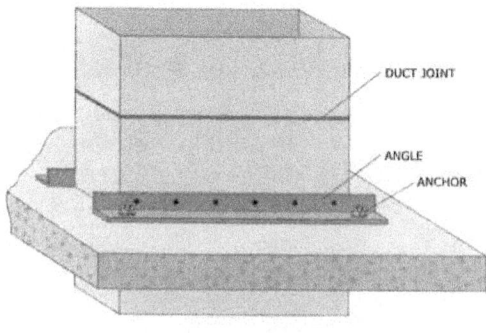

DUCT JOINT

ANGLE

ANCHOR

RIGID CONNECTION

Figure 59: Penetration for interior rectangular duct.

 Use the shortest screws possible when penetrating ductwork to minimize airflow noise inside the duct.

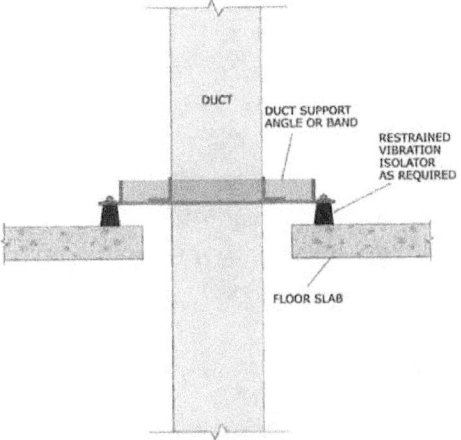

DUCT

DUCT SUPPORT
ANGLE OR BAND

RESTRAINED
VIBRATION
ISOLATOR
AS REQUIRED

FLOOR SLAB

Figure 60: Penetration for interior round duct.

Step 1: Lay out the location of penetration

Coordinate the layout with a structural engineer. Additional structural supports may be required.

Lay out all attachment points before anchoring.

Step 2: Install anchors for bracing

For instructions on installing anchors, see Anchors (page 107).

Step 3: Run duct as required by approved construction documents

Attach duct to the angle assembly with angles and sheet metal screws. Attach angles to the building structure.

 Use the shortest screws possible when penetrating ductwork to minimize airflow noise inside the duct.

 Duct penetrations may be similar to pipe penetrations (page 79).

END OF DETAIL.

PIPE BRACING DETAILS AND INSTALLATION INSTRUCTIONS

This section gives instructions on bracing five different kinds of piping:

- Suspended piping (this page).
- Floor-mounted piping (page 70).
- Roof-mounted piping (page 73).
- Wall-mounted piping (page 76).
- Pipe penetrations (page 79).

Suspended Piping

The seven ways to brace suspended piping are by using:

- Clevis hanger braced at the restraining bolt (page 55).
- Clevis hanger braced with cables (page 57).
- Clevis hanger braced at the hanger rod (page 58).
- Pipe clamps (page 60).
- Isolated pipe with clevis hanger (page 65).
- Trapeze support system (page 66).
- Double roller for expansive pipe (page 68).

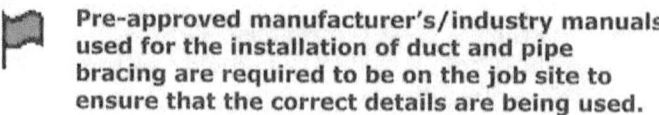 **For post-tension (pre-stressed) buildings, locate the tendons before drilling. Extreme damage may occur if a tendon is nicked or cut.**

Refer to approved construction documents for details and provisions for crossing fire barriers, area separation walls/floors/roofs, smoke barriers, and seismic separation joints.

Pre-approved manufacturer's/industry manuals used for the installation of duct and pipe bracing are required to be on the job site to ensure that the correct details are being used.

Suspended using clevis hanger braced at the restraining bolt

Step 1: Attach vertical rods with hanger to the building structure

Lay out all attachment points before anchoring, then refer to Attachment Details Connecting to Building Structure (page 102). For instructions on installing anchors, see Anchors (page 107).

 Building structure must be point-load capable. Verify with the appropriate design professional.

Step 2: Run pipe as required by approved construction documents

Step 3: Install anchors for bracing

For instructions on installing anchors, see Anchors (page 107).

**Bracing Details and Installation Instructions:
Suspended Piping**

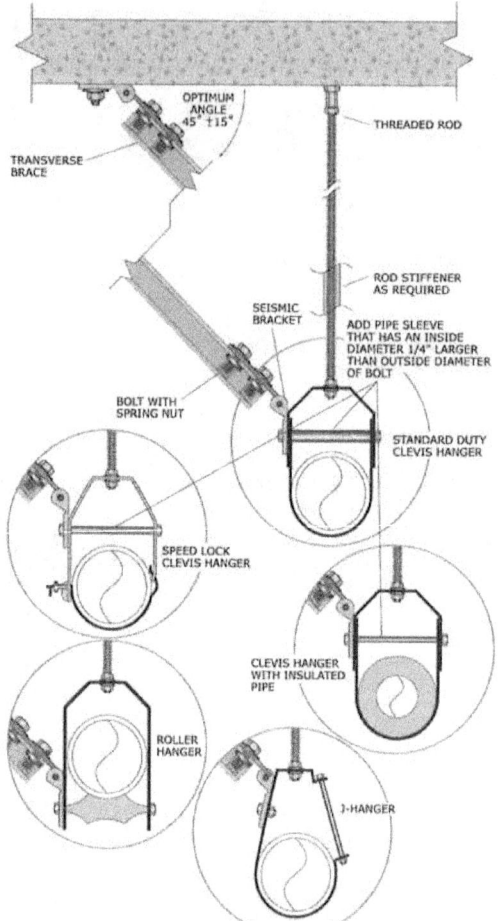

Figure 61: Single clevis hanger support with strut or angle
transverse bracing at the restraining bolt.

 **Pre-approved manufacturer's/industry
manuals may limit the maximum transverse
and longitudinal angles to 45 degrees.**

 **Torque bolts per manufacturer's
recommendations.**

END OF DETAIL.

Suspended using clevis hanger braced with cables

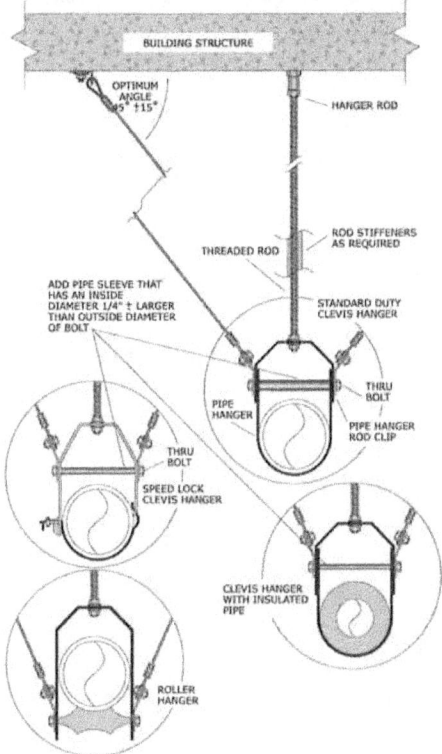

Figure 62: Single clevis hanger support with cable transverse bracing at the restraining bolt.

For cable assembly instructions, see Cables (page 134). For details on attaching cable to the building structure, refer to Attachment Details Connecting to Building Structure (page 102). For angle and strut attachment, see Figure 61 (page 56).

 Pre-approved manufacturer's/industry manuals may limit the maximum transverse and longitudinal angles to 45 degrees.

 Torque bolts per manufacturer's recommendations.

END OF DETAIL.

Suspended using clevis hanger braced at the hanger rod

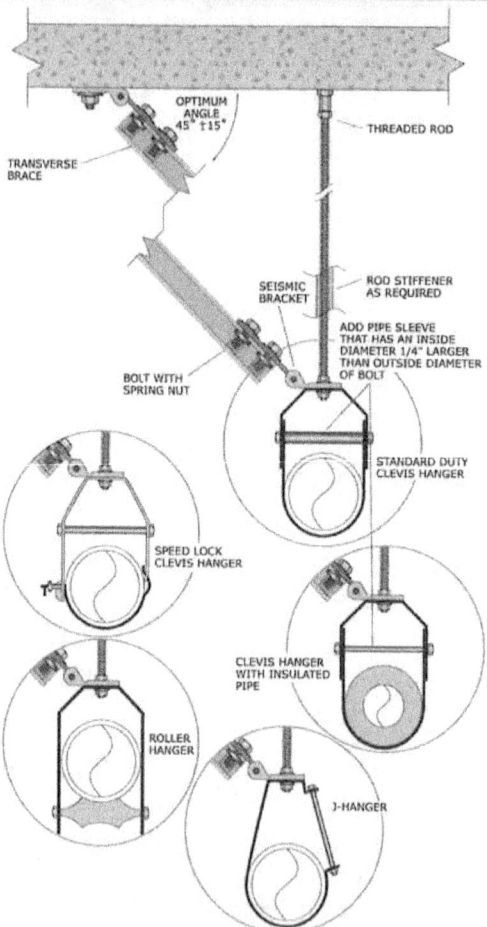

Figure 63: Single clevis hanger support with strut or angle transverse bracing at hanger rod.

 Pre-approved manufacturer's/industry manuals may limit the maximum transverse and longitudinal angles to 45 degrees.

 Torque bolts per manufacturer's recommendations.

Step 1: Attach vertical rods with hangers to the building structure

Lay out all attachment points before anchoring, then refer to Attachment Details Connecting to Building Structure (page 102). For instructions on installing anchors, see Anchors (page 107).

 Building structure must be point-load capable. Verify with the appropriate design professional.

Step 2: Run pipe as required by approved construction documents

Step 3: Install anchors for bracing

For cable assembly instructions, see Cables (page 134). For details on attaching cable to the building structure, see Attachment Details Connecting to Building Structure (page 102). For angle and strut attachment, see Figure 63 (page 58). For instructions on installing anchors, see Anchors (page 107).

 Insulate pipe with weatherproof jacket where required.

END OF DETAIL.

Bracing Details and Installation Instructions:
Suspended Piping

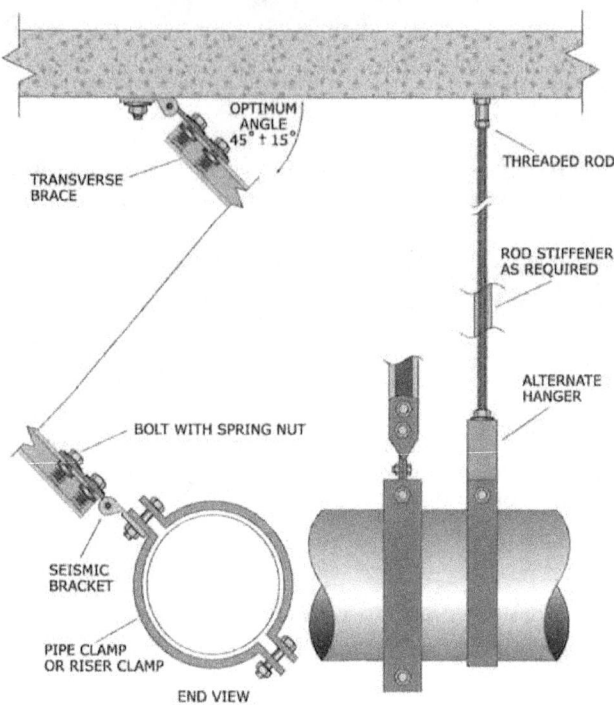

Figure 64: Pipe clamp supports with transverse strut or angle, transverse brace and hanger rod.

 Pre-approved manufacturer's/industry manuals may limit the maximum transverse and longitudinal angles to 45 degrees.

 Torque bolts per manufacturer's recommendations.

 Insulate after attaching pipe clamp.

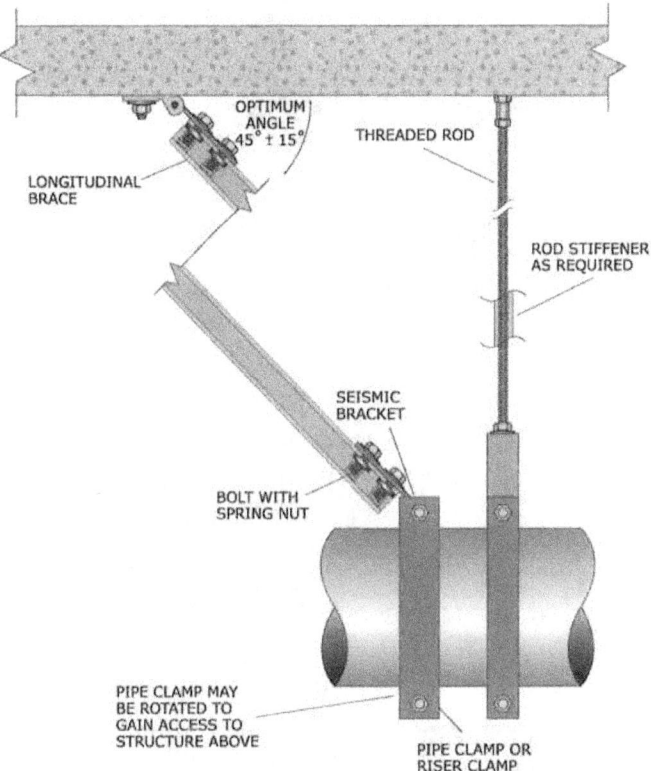

Figure 65: Pipe clamp supports with longitudinal strut or angle, longitudinal brace and hanger rod.

 Pre-approved manufacturer's/industry manuals may limit the maximum transverse and longitudinal angles to 45 degrees.

 Torque bolts per manufacturer's recommendations.

 Insulate after attaching pipe clamp.

Bracing Details and Installation Instructions:
Suspended Piping

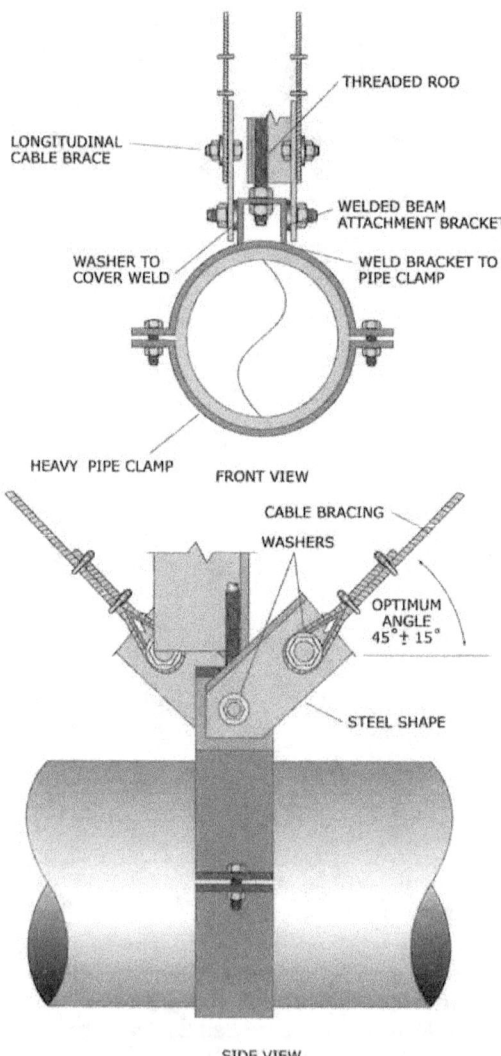

THREADED ROD

LONGITUDINAL
CABLE BRACE

WELDED BEAM
ATTACHMENT BRACKET

WASHER TO
COVER WELD

WELD BRACKET TO
PIPE CLAMP

HEAVY PIPE CLAMP

FRONT VIEW

CABLE BRACING

WASHERS

OPTIMUM
ANGLE
45° ± 15°

STEEL SHAPE

SIDE VIEW

Figure 66: Pipe clamp supports with longitudinal cable brace and hanger rod.

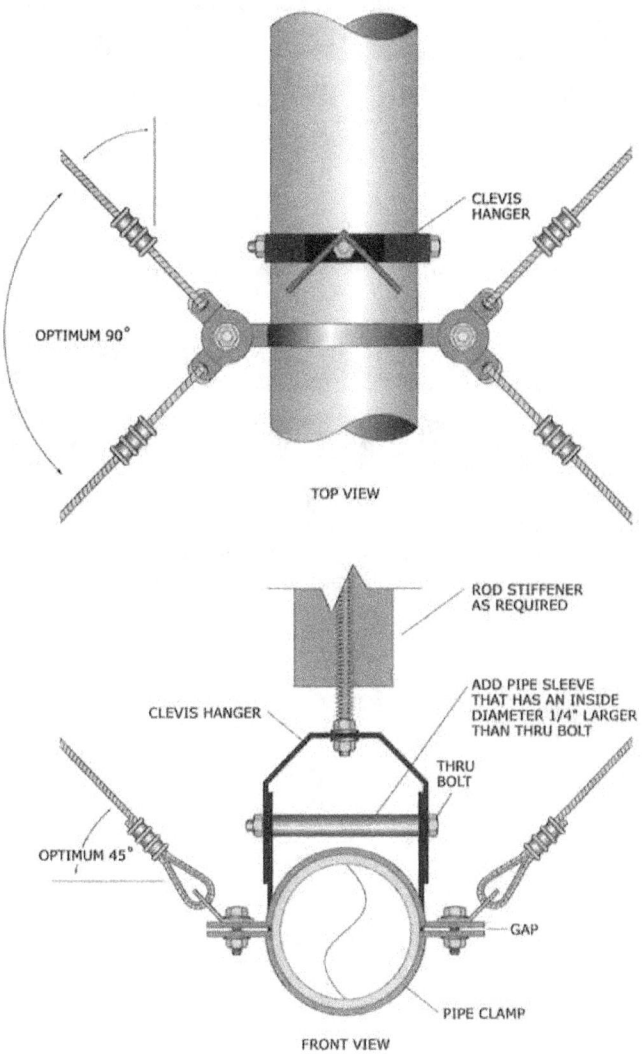

CLEVIS
HANGER

OPTIMUM 90°

TOP VIEW

ROD STIFFENER
AS REQUIRED

ADD PIPE SLEEVE
THAT HAS AN INSIDE
DIAMETER 1/4" LARGER
THAN THRU BOLT

CLEVIS HANGER

THRU
BOLT

OPTIMUM 45°

GAP

PIPE CLAMP

FRONT VIEW

Figure 67: Pipe clamp supports with all directional brace and hanger rod.

Bracing Details and Installation Instructions:
Suspended Piping

Step 1: Attach vertical rods with hanger or vibration isolator (as required) to the building structure

Lay out all attachment points before anchoring, then refer to Attachment Details Connecting to Building Structure (page 102). For instructions on installing anchors, see Anchors (page 107).

 Building structure must be point load capable. Verify with the appropriate design professional.

Step 2: Run pipe as required by approved construction documents

Step 3: Install anchors for bracing

For cable assembly instructions, see Cables (page 134). For details on attaching cable to the building structure, refer to Attachment Details Connecting to Building Structure (page 102). For angle and strut attachment, see Figure 69 (page 66). For instructions on installing anchors, see Anchors (page 107).

END OF DETAIL.

Vibration-isolated pipe with clevis hanger

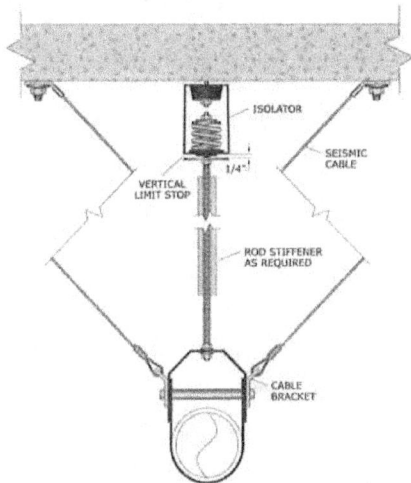

Figure 68: Vibration-isolated single pipe.

Step 1: Attach vertical rods with hanger or vibration isolator (as required) to the building structure

Lay out all attachment points before anchoring, then refer to Attachment Details Connecting to Building Structure (page 102). For instructions on installing anchors, see Anchors (page 107).

 Building structure must be point load capable. Verify with the appropriate design professional.

Step 2: Run pipe as required by approved construction documents

Step 3: Install anchors for bracing

For cable assembly instructions, see Cables (page 134). For details on attaching cable to the building structure, refer to Attachment Details Connecting to Building Structure (page 102). For instructions on installing anchors, see Anchors (page 107).

END OF DETAIL.

Bracing Details and Installation Instructions:
Suspended Piping

Suspended with trapeze support system

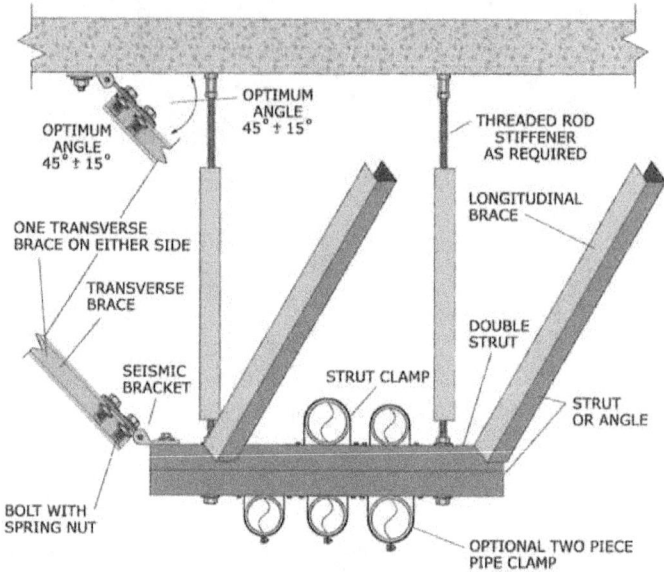

Figure 69: Trapeze support with strut or angle lateral supports.

 Pre-approved manufacturer's/industry manuals may limit the maximum transverse and longitudinal angles to 45 degrees.

 Separate isolated piping from rigidly braced piping.

Step 1: Attach vertical rods with hanger to the building structure

Lay out all attachment points before anchoring, then refer to Attachment Details Connecting to Building Structure (page 102). For instructions on installing anchors, see Anchors (page 107).

 Building structure must be point load capable. Verify with the appropriate design professional.

Step 2: Run pipe as required by approved construction documents

Step 3: Install anchors for bracing

For cable assembly instructions, see Cables (page 134). For details on attaching cable to the building structure, refer to Attachment Details Connecting to Building Structure (page 102). For angle and strut attachment, see Figure 69 (page 66). For instructions on installing anchors, see Anchors (page 107).

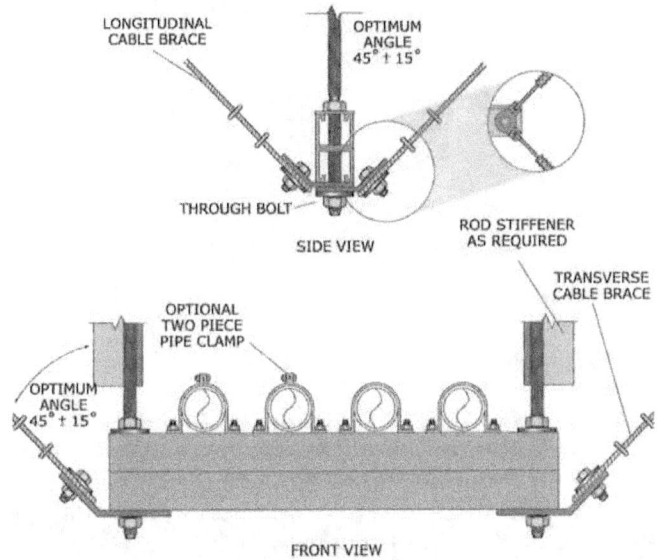

Figure 70: Trapeze support with cable lateral brace.

 Separate isolated piping from rigidly braced piping.

END OF DETAIL.

Suspended with double roller support system

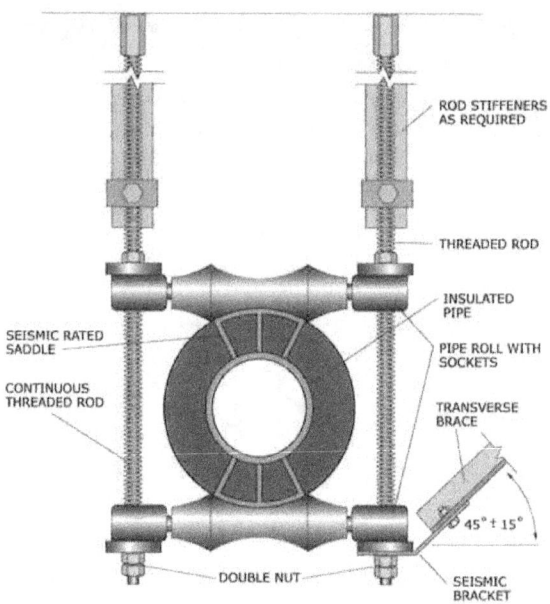

ROD STIFFENERS
AS REQUIRED

THREADED ROD

INSULATED
PIPE

SEISMIC RATED
SADDLE

PIPE ROLL WITH
SOCKETS

CONTINUOUS
THREADED ROD

TRANSVERSE
BRACE

45° ± 15°

DOUBLE NUT

SEISMIC
BRACKET

Figure 71: Double roller support for thermally expansive piping.

 Pre-approved manufacturer's/industry manuals may limit the maximum transverse and longitudinal angles to 45 degrees.

Step 1: Attach vertical rods with hanger to the building structure

Lay out all the attachment points before anchoring, then refer to Attachment Details Connecting to Building Structure (page 102). For instructions on installing anchors, see Anchors (page 107).

 Building structure must be point-load capable. Verify with the appropriate design professional.

Step 2: Run pipe as required by approved construction documents

Step 3: Install anchors for bracing

For cable assembly instructions, see Cables (page 134). For angle and strut attachment, see Figure 69 (page 66). For instructions on installing anchors, see Anchors (page 107).

END OF DETAIL.

Floor-mounted Piping

Attach to the floor with angles either in a single pipe support configuration as shown in Figure 72 (below) or on a trapeze as shown in Figure 73 (page 71) and Figure 74 (page 72).

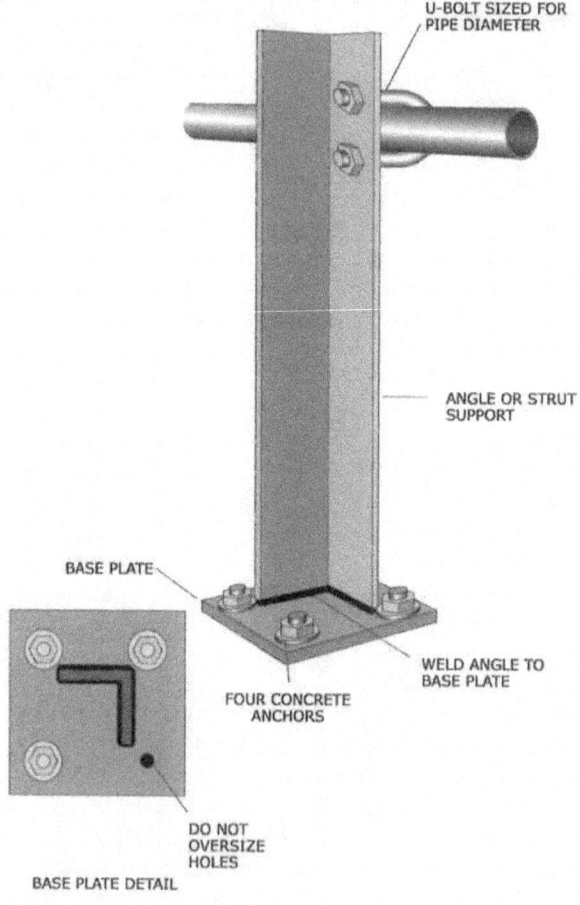

U-BOLT SIZED FOR PIPE DIAMETER

ANGLE OR STRUT SUPPORT

BASE PLATE

WELD ANGLE TO BASE PLATE

FOUR CONCRETE ANCHORS

DO NOT OVERSIZE HOLES

BASE PLATE DETAIL

Figure 72: Single vertical support.

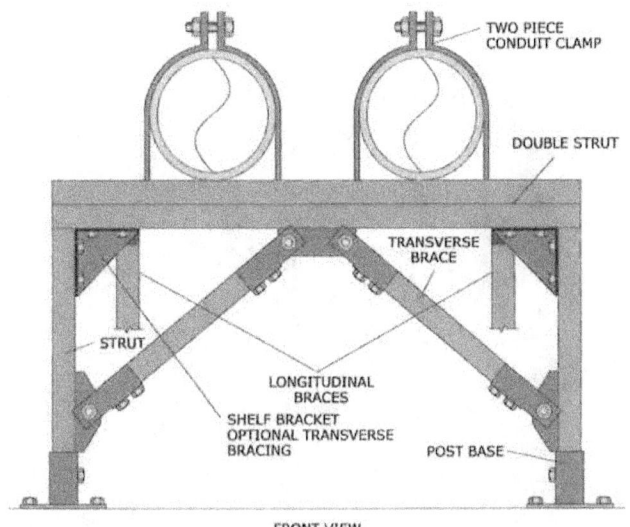

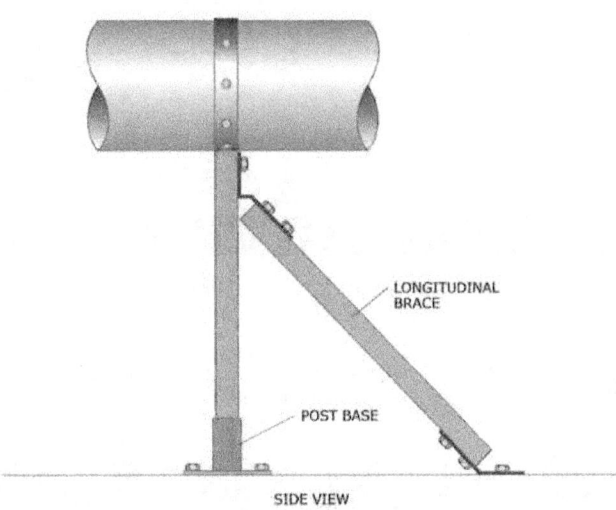

Figure 73: Attachment to a floor with strut trapeze.

**Bracing Details and Installation Instructions:
Floor-mounted Piping**

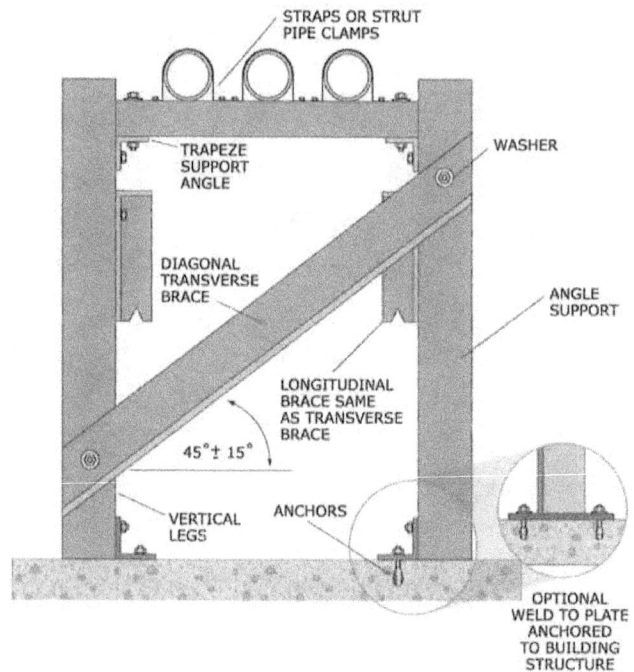

Figure 74: Attachment to a floor with steel shaped trapeze.

Step 1: Attach supports or angles to the floor

For instructions on installing anchors, see Anchors (page 107).

 Building structure must be point-load capable. Verify with the appropriate design professional.

Step 2: Build trapeze support and attach piping to support with straps or strut pipe clamps

 Torque bolts per manufacturer's recommendations.

END OF DETAIL.

Roof-mounted Piping

The three ways of attaching piping to a roof are:

- Single pipe support (this page).
- Wood blocking support (page 74).
- Trapeze support (page 75).

Single pipe support

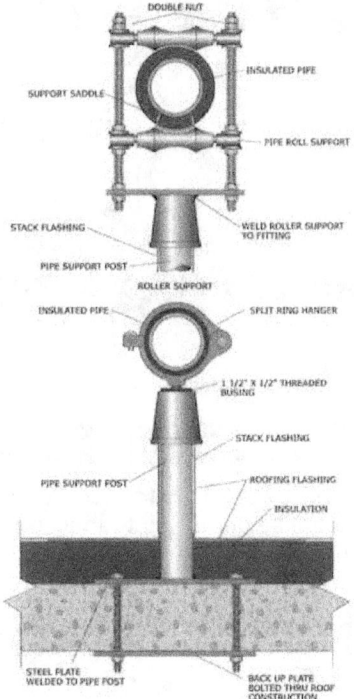

Figure 75: Single pipe support.

Step 1: Attach vertical support to the building structure

Lay out all attachment points before anchoring, then refer to Attachment Details Connecting to Building Structure (page 102).For instructions on installing anchors, see Anchors (page 107).

Step 2: Assemble support and attach to building structure

 Apply flashing to roof penetration.

Step 3: Run pipe as required by approved construction documents

END OF DETAIL.

Wood blocking support

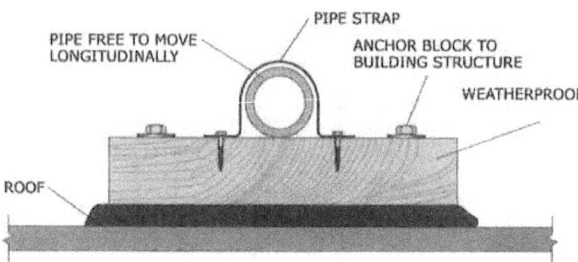

Figure 76: Wood blocking support.

Step 1: Attach wood support to the building structure

Lay out all attachment points before anchoring, then refer to Attachment Details Connecting to Building Structure (page 102). For instructions on installing anchors, see Anchors (page 107).

Step 2: Apply flashing or sealant to roof penetration

 Follow the manufacturer's instructions.

Step 3: Run pipe as required by approved construction documents

END OF DETAIL.

Trapeze support

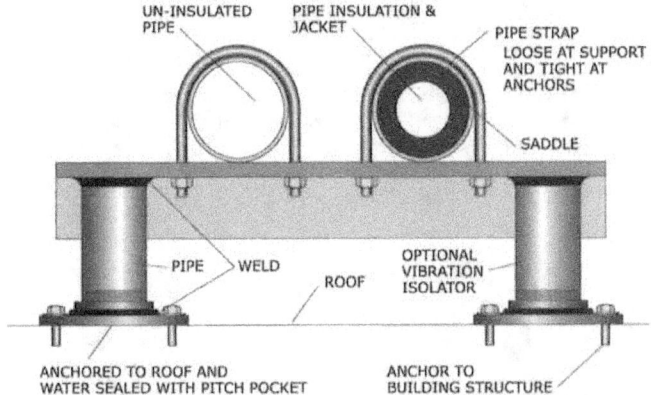

UN-INSULATED
PIPE

PIPE INSULATION &
JACKET

PIPE STRAP
LOOSE AT SUPPORT
AND TIGHT AT
ANCHORS

SADDLE

PIPE   WELD

ROOF

OPTIONAL
VIBRATION
ISOLATOR

ANCHORED TO ROOF AND
WATER SEALED WITH PITCH POCKET

ANCHOR TO
BUILDING STRUCTURE

Figure 77: Trapeze support.

 Refer to approved contract documents for pitch pocket details.

 Separate isolated piping from rigidly braced piping.

Step 1: Lay out the pipe run

Lay out all attachment points before anchoring.

Step 2: Install anchors for bracing

For instructions on installing anchors, see Anchors (page 107).

Step 3: Assemble angles and secure to anchors

Step 4: Run pipe as required by approved construction documents

Attach piping to trapeze support with pipe straps or U-bolts. Provide hard insulation and sheet metal protection shield at the support area.

END OF DETAIL.

Wall-mounted Piping

Directly attach to the wall with two-hole pipe clamps as shown in Figure 78 (below) or with angle brackets as shown in Figure 79 (page 77) and Figure 80 (page 78).

Piping surface-mounted from the underside of a building structural slab or rated structural ceiling should be attached as shown in Figure 78 (below).

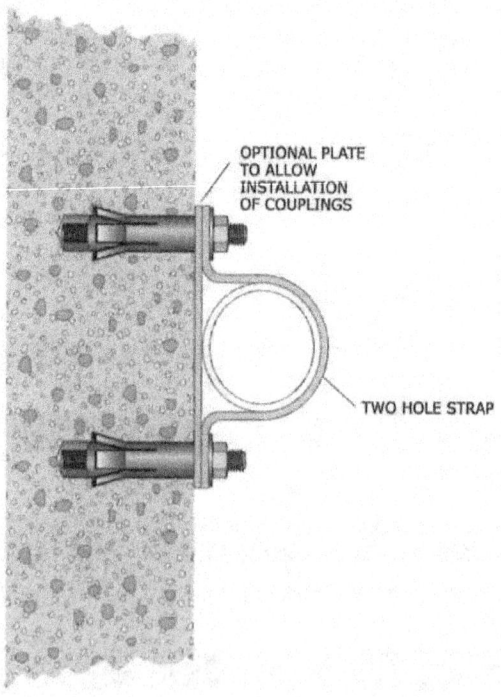

OPTIONAL PLATE
TO ALLOW
INSTALLATION
OF COUPLINGS

TWO HOLE STRAP

Figure 78: Direct attachment.

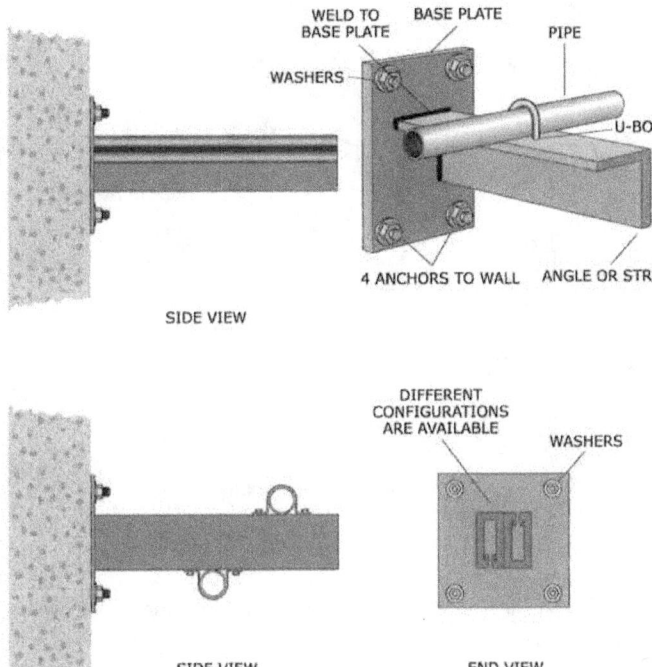

Figure 79: Attachment to the wall with angle or strut welded to attachment plate.

 Refer to approved construction documents for limitations to length of standoffs.

**Bracing Details and Installation Instructions:
Wall-mounted Piping**

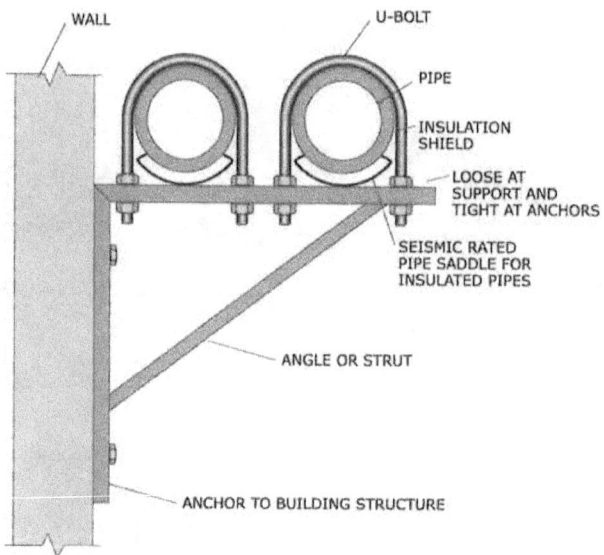

Figure 80: Attachment to studs in the wall with pre-manufactured brackets.

Step 1: If required, attach supports or angles to the wall

For instructions on installing anchors, see Anchors (page 107).

 Building structure must be point-load capable. Verify with the appropriate design professional.

For drywall attachments, use a strut attachment to the studs as shown in Figure 55 (page 48).

Step 2: Attach pipe to support with straps

END OF DETAIL.

Pipe Penetrations

The two types of pipe penetrations are:

- Roof pipe penetrations (below).
- Interior pipe penetrations (page 82).

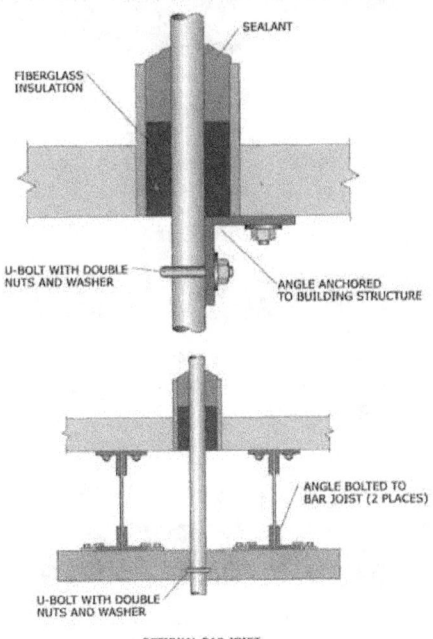

Figure 81: Roof pipe penetration detail.

 Refer to Attachment Details Connecting to Building Structure (page 102) for bar joist attachment.

 Verify that the bar joist is point load capable.

**Bracing Details and Installation Instructions:
Pipe Penetrations**

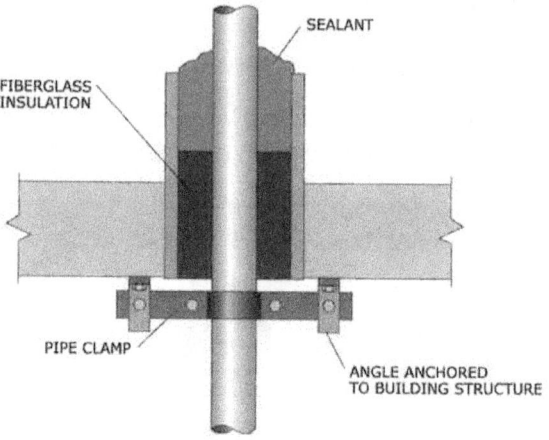

Figure 82: Roof pipe penetration detail.

Step 1: Lay out location of penetration

 Coordinate the layout with a structural engineer. Additional structural supports may be required.

Lay out all attachment points before anchoring.

Step 2: Install anchors and attach angles to building structure

For instructions on installing anchors, see Anchors (page 107).

Step 3: Run pipe as required by approved construction documents

Attach pipe to angle assembly with angles and U-bolts or with pipe clamp.

 Add flashing as required by approved construction documents or as directed in manufacturer's instructions

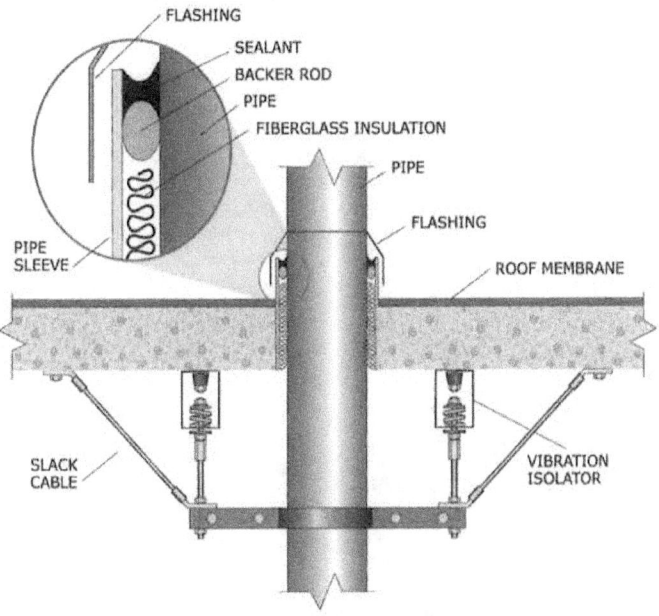

Figure 83: Isolated pipe roof penetration detail.

 Separate isolated piping from rigidly braced piping.

END OF DETAIL.

Interior Pipe Penetrations

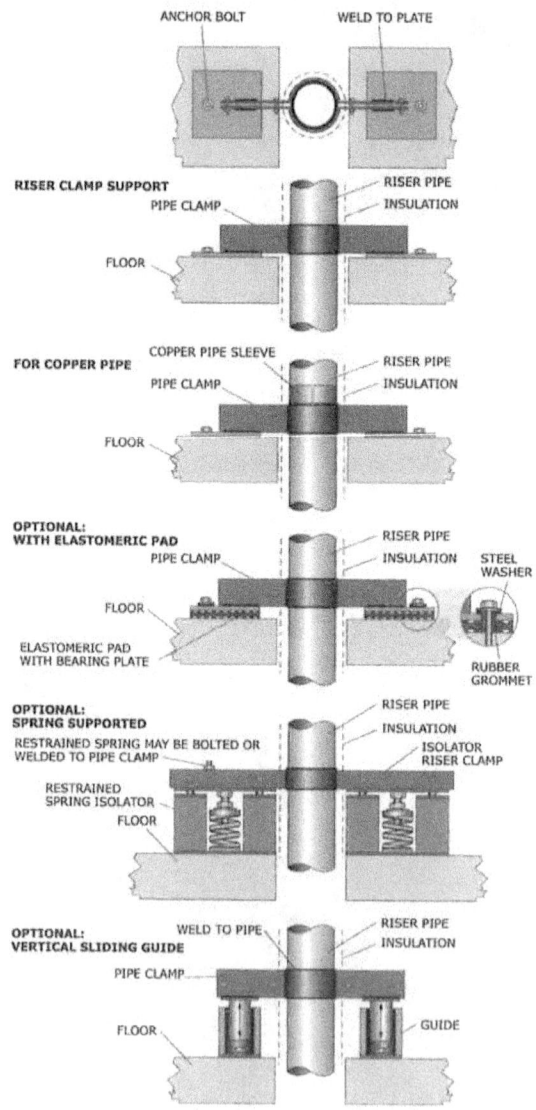

Figure 84: Interior pipe penetration detail.

Step 1: Lay out location of penetration

 Coordinate the layout with a structural engineer. Additional structural supports may be required.

Lay out all attachment points before anchoring.

Step 2: Install anchors and attach angles to building structure

For instructions on installing anchors, see Anchors (page 107).

Step 3: Run pipe as required by approved construction documents

Attach pipe to angle assembly with angles and U-bolts or with pipe clamp.

 Piping penetrating walls/floor slabs/roofs must be installed per approved construction documents, submittals, or manufacturer's instructions.

 Rigid attachment to lightweight walls may cause vibration problems.

 Separate isolated piping from rigidly braced piping.

END OF DETAIL.

SUSPENDED EQUIPMENT ATTACHMENT

 Do not mix bracing systems for strut and cable bracing.

This section provides instructions for equipment that is:

* Suspended by threaded rods connected to equipment brackets or additional steel supports (this page).

* Suspended by steel shapes (page 95).

* Suspended by threaded rods for in-line pipe equipment (page 97).

Suspended by Threaded Rods Connected to Equipment Brackets or Additional Steel Supports

The four ways to suspend equipment with threaded rods or steel supports are:

* Rigid connection to the building structure using four threaded rods with lateral cable bracing (page 85).

* Rigid connection to the building structure using four threaded rods with lateral steel shaped bracing (page 88).

* Vibration-isolated connection to the building structure using a minimum of four threaded rods and lateral cable bracing (page 90).

* Two-point equipment attachment—bolted to the building structure (page 93).

**Rigid connection to the building structure using four
threaded rods with lateral cable bracing**

 **Equipment should have pre-installed brackets
that can support the attachment to the
building.**

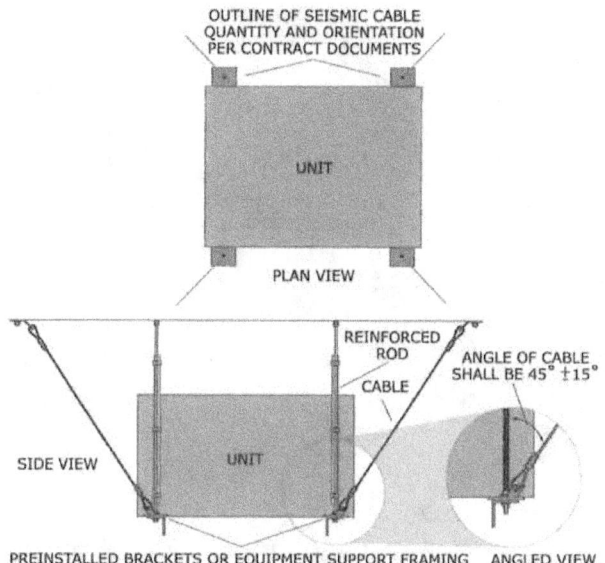

Figure 85: Rigid connection to the building structure.

 **Cables provide horizontal support for seismic
loads and should not be installed to hang
equipment.**

**Step 1: Attach the equipment to the building structure
using threaded rods and anchors**

Lay out all attachment points before anchoring, then refer
to Attachment Details Connecting to Building Structure
(page 102). For instructions on installing anchors, see
Anchors (page 107).

Suspended Equipment: Threaded Rods Connected to Equipment Brackets or Additional Steel Supports

Step 2: Add rod stiffeners

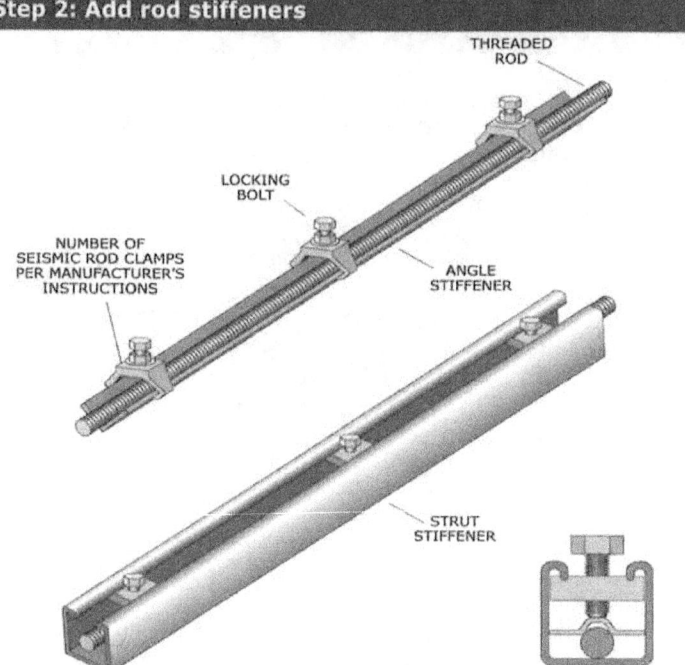

THREADED ROD

LOCKING BOLT

NUMBER OF SEISMIC ROD CLAMPS PER MANUFACTURER'S INSTRUCTIONS

ANGLE STIFFENER

STRUT STIFFENER

Figure 86: Rod stiffeners.

Step 3: Install anchors for cable attachment

Lay out all attachment points before anchoring, then refer to Attachment Details Connecting to Building Structure (page 102). For instructions on installing anchors, see Anchors (page 107).

Step 4: Attach cable to the building structure

For cable assembly instructions, see Cables (page 134). For details on attaching cable to the building structure, refer to Attachment Details Connecting to Building Structure (page 102).

Step 5: Attach cables to equipment

For details on attaching cable to the equipment, see Figure
87 (below).

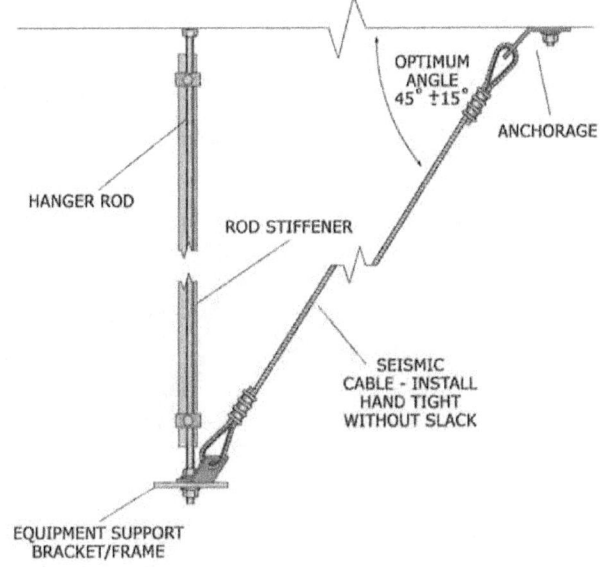

OPTIMUM
ANGLE
45° ±15°

ANCHORAGE

HANGER ROD

ROD STIFFENER

SEISMIC
CABLE - INSTALL
HAND TIGHT
WITHOUT SLACK

EQUIPMENT SUPPORT
BRACKET/FRAME

Figure 87: Attachment of cable to the equipment.

END OF DETAIL.

Suspended Equipment: Threaded Rods Connected to Equipment Brackets or Additional Steel Supports

Rigid connection to the building structure using four threaded rods with lateral steel shaped bracing

Equipment may have pre-installed brackets for angle support attachments as shown in Figure 88 (below).

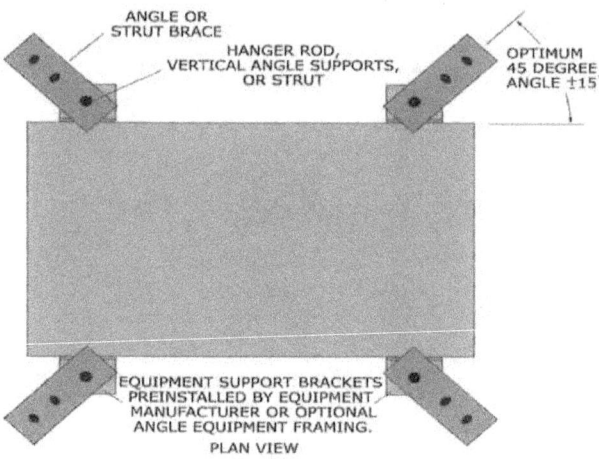

Figure 88: Rigid attachment of angles to the building structure.

Step 1: Attach the equipment to the building structure using threaded rods and anchors

Lay out all attachment points before anchoring, then refer to Attachment Details Connecting to Building Structure (page 102). For instructions on installing anchors, see Anchors (page 107). Rod stiffeners are not required.

Step 2: Install anchors for angle or strut supports

For building structure attachment details, see Attachment Details Connecting to Building Structure (page 102). For instructions on installing anchors, see Anchors (page 107).

Suspended Equipment: Threaded Rods Connected to Equipment Brackets or Additional Steel Supports

Step 3: Attach angles or strut supports to the building structure

Step 4: Attach angles or struts to equipment

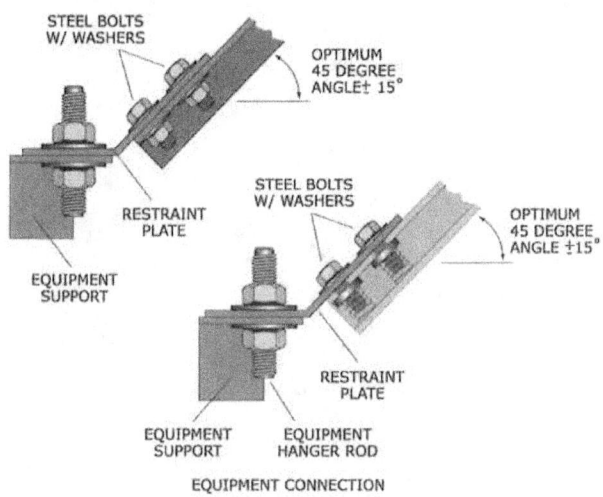

Figure 89: Attachment of angle or strut to the equipment.

END OF DETAIL.

**Vibration-isolated connection to the building structure
using a minimum of four threaded rods and lateral
cable bracing**

Equipment may have pre-installed brackets for angle
support attachments. See Figure 90 (below) and Figure 91
(below).

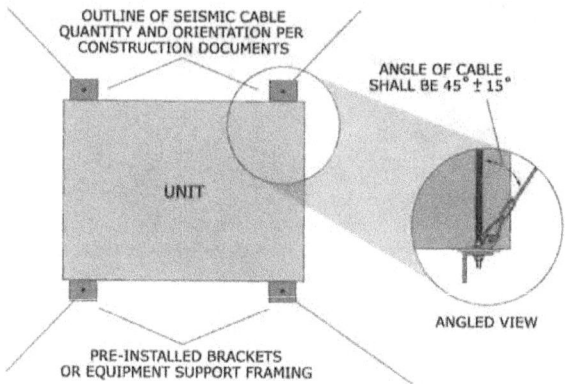

Figure 90: Plan view of vibration-isolated, suspended attachment to
the building structure.

Side view shows vibration isolators, rods (without rod
stiffeners), and cables.

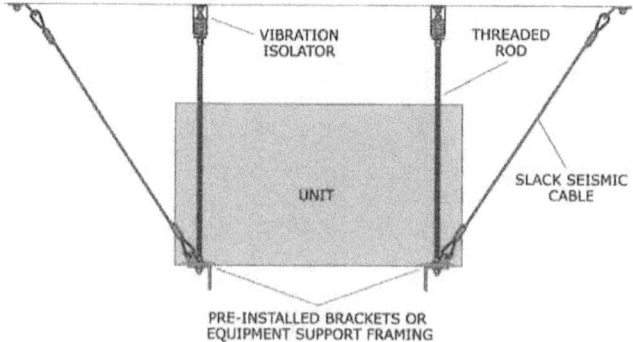

Figure 91: Side view of vibration-isolated, suspended attachment
to the building structure.

Step 1: Attach equipment to the building structure using threaded rods, isolators and anchors

For isolator detail, see Figure 92 (below). For building structure attachment details, refer to Attachment Details Connecting to Building Structure (page 102).

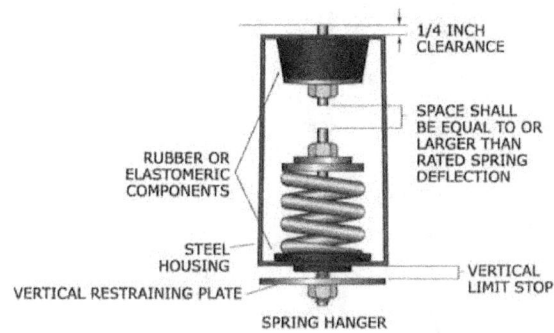

Figure 92: Isolator detail.

Step 2: Install anchors and attach cable to the building structure

For cable connection to the building structure, refer to Attachment Details Connecting to Building Structure (page 102). For instructions on installing anchors, see Anchors (page 107).

Step 3: Attach cables to equipment

For cable assembly, see Cables (page 134). For cable
attachment to equipment, see Figure 93 (below).

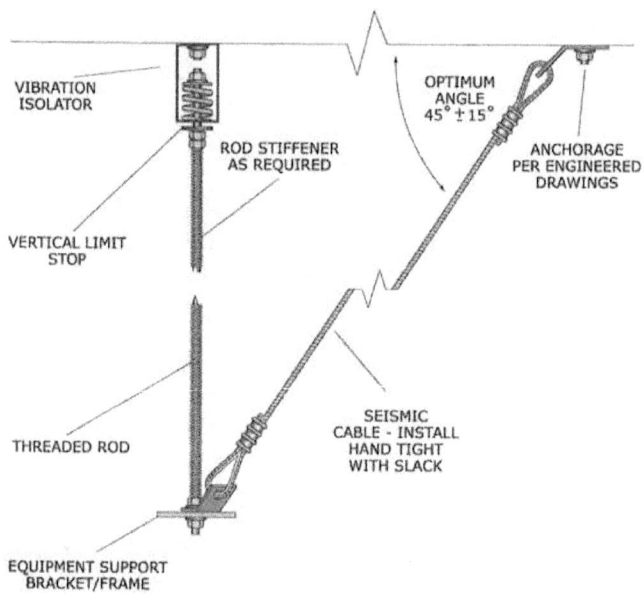

VIBRATION
ISOLATOR

OPTIMUM
ANGLE
45° ±15°

ROD STIFFENER
AS REQUIRED

ANCHORAGE
PER ENGINEERED
DRAWINGS

VERTICAL LIMIT
STOP

THREADED ROD

SEISMIC
CABLE - INSTALL
HAND TIGHT
WITH SLACK

EQUIPMENT SUPPORT
BRACKET/FRAME

Figure 93: Attachment of cable/rod assembly to the equipment.

Step 4: Re-adjust vertical stop limit after support is fully loaded

Verify the adjustments of the vibration isolators with
manufacturer's/industry manuals.

END OF DETAIL.

Suspended Equipment: Threaded Rods Connected to Equipment Brackets or Additional Steel Supports

Two-point equipment attachment—bolted to the building structure

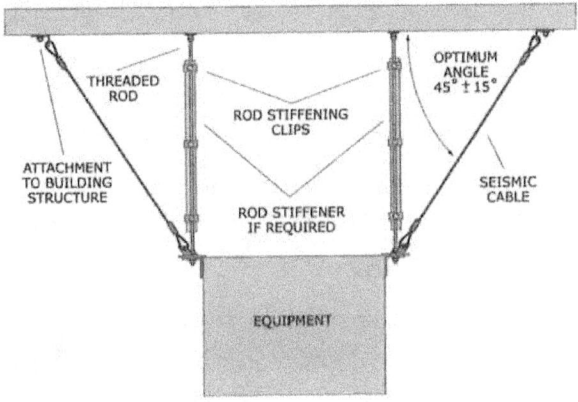

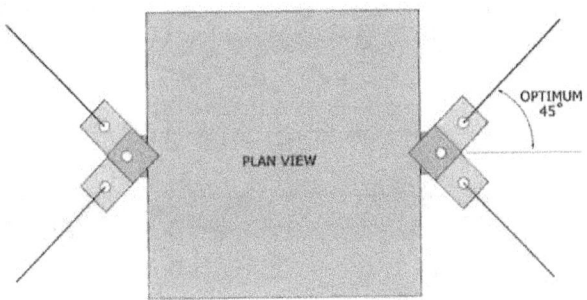

Figure 94: Two-point equipment attachment.

Suspended Equipment: Threaded Rods Connected to Equipment Brackets or Additional Steel Supports

Step 1: Attach anchors and vertical rods to the building structure

Lay out all attachment points before anchoring, then refer to Attachment Details Connecting to Building Structure (page 102). Attach equipment to the vertical rods.

 The equipment attachment should be located just above the center of gravity of the equipment to minimize swinging. It should be a rigid attachment with brackets to the equipment using double nuts and washers, especially if connected at the top as shown in Figure 94 (page 93).

Step 2: Attach rod stiffeners

For attachment details, refer to Figure 86 (page 86).

Step 3: Install anchors for cable attachment

For typical anchorage to different building construction, refer to Attachment Details Connecting to Building Structure (page 102). For details on bolting directly to building structure, see Anchors (page 107).

Step 4: Attach cables to the building structure

For cable assembly see Cables (page 134). For details on attaching cable to the building structure, refer to Attachment Details Connecting to Building Structure (page 102).

Step 5: Attach cables to equipment

The detail in Figure 87 (page 87) shows the cable attachment to the equipment.

END OF DETAIL.

Suspended by Steel Shapes

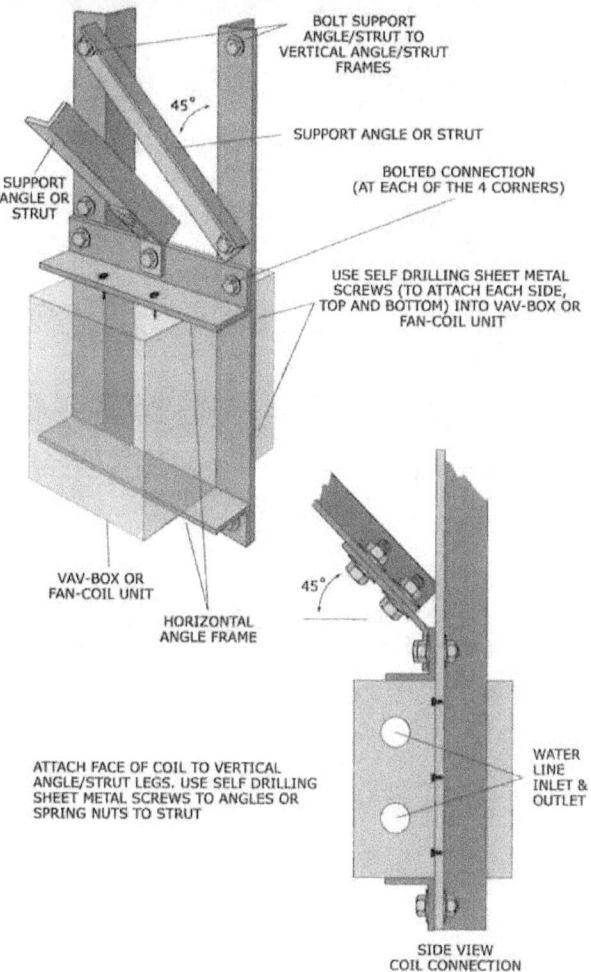

BOLT SUPPORT
ANGLE/STRUT TO
VERTICAL ANGLE/STRUT
FRAMES

45°

SUPPORT ANGLE OR STRUT

BOLTED CONNECTION
(AT EACH OF THE 4 CORNERS)

SUPPORT
ANGLE OR
STRUT

USE SELF DRILLING SHEET METAL
SCREWS (TO ATTACH EACH SIDE,
TOP AND BOTTOM) INTO VAV-BOX OR
FAN-COIL UNIT

VAV-BOX OR
FAN-COIL UNIT

45°

HORIZONTAL
ANGLE FRAME

WATER
LINE
INLET &
OUTLET

ATTACH FACE OF COIL TO VERTICAL
ANGLE/STRUT LEGS. USE SELF DRILLING
SHEET METAL SCREWS TO ANGLES OR
SPRING NUTS TO STRUT

SIDE VIEW
COIL CONNECTION

Figure 95: Attachment of double angles for equipment support.

Use this type of installation for duct-mounted coils, VAV boxes, or fan-coil units weighing less than 150 pounds.

Suspended Equipment: Steel Shapes

Step 1: Attach anchors and vertical angles or strut to the building structure

For building structure attachment details, refer to Attachment Details Connecting to Building Structure (page 102).

Step 2: Attach horizontal framing

For attachment details, refer to Attachment Details Connecting to Building Structure (page 102).

Step 3: Install anchors for angle or strut restraints

For typical anchorage to different building construction, refer to Attachment Details Connecting to Building Structure (page 102). For instructions on installing anchors, see Anchors (page 107).

Step 4: Attach support angles or struts

One support is attached to the two vertical angles or struts. One support is attached to the building structure and to the top horizontal frame. For details on angle or strut attachment, refer to Attachment Details Connecting to Building Structure (page 102).

Step 5: Attach equipment

Attach equipment to the support assembly as shown in Figure 95 (page 95).

END OF DETAIL.

Suspended by Threaded Rods for In-line Pipe Equipment

This section decribes the following three types of in-line pipe equipment suspended by threaded rods:

- In-line pump (below).
- In-line air separator (page 99).
- In-line heat exchanger (page 100).

In-line Pump

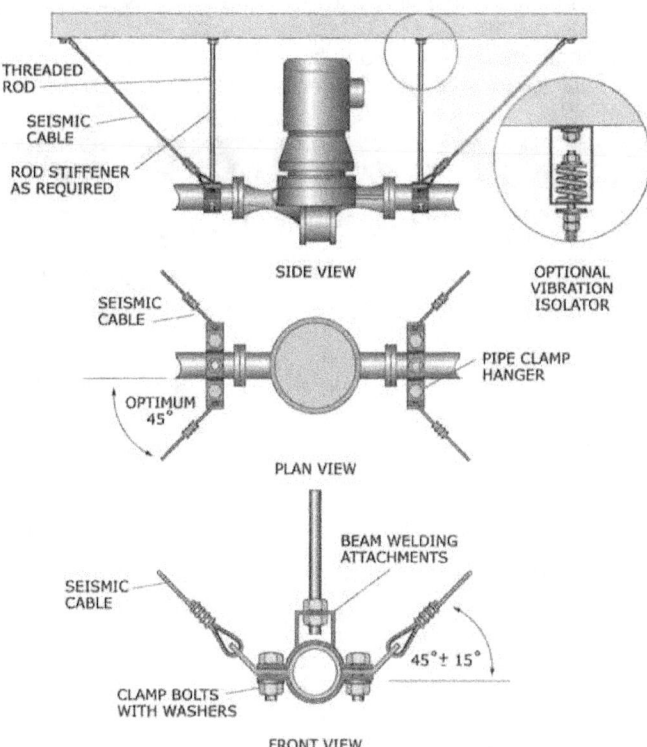

Figure 96: In-line pump.

**Suspended Equipment: Threaded Rods
for In-line Pipe Equipment**

Step 1: Attach pipe clamps to the building structure using threaded rods and anchors

Lay out all attachment points before anchoring, then refer to Attachment Details Connecting to Building Structure (page 102). For instructions on installing anchors, see Anchors (page 107).

Step 2: Add rod stiffeners

For rod stiffener details, refer to Figure 86 (page 86).

Step 3: Install anchors for cable bracing

Attach cable seismic brackets with anchors. For typical anchorage to different building construction, refer to Attachment Details Connecting to Building Structure (page 102). For instructions on installing anchors, see Anchors (page 107).

Step 4: Attach seismic cable

For cable assembly, see Cables (page 134).

END OF DETAIL.

In-line Air Separator

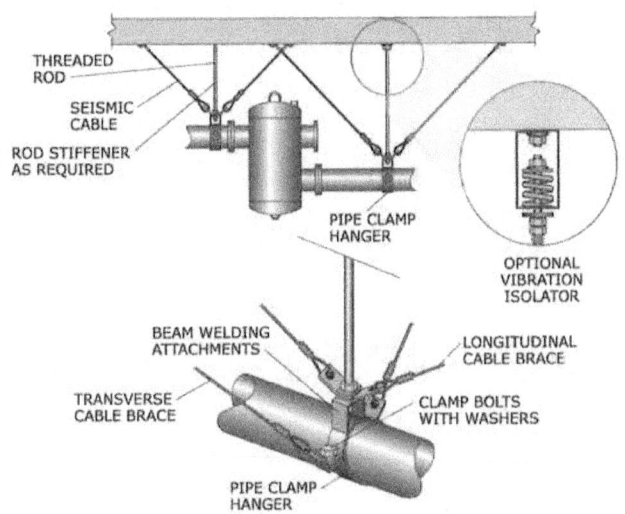

Figure 97: In-line air separator.

Step 1: Attach pipe clamps to the building structure using threaded rods and anchors

Lay out all attachment points before anchoring, then refer to Attachment Details Connecting to Building Structure (page 102). For instructions on installing anchors, see Anchors (page 107).

Step 2: Add rod stiffeners

For rod stiffener details, refer to Figure 86 (page 86).

Step 3: Install anchors for cable bracing

Attach cable seismic brackets with anchors. For typical
anchorage to different building construction, refer to
Attachment Details Connecting to Building Structure (page
102). For instructions on installing anchors, see Anchors
(page 107).

Step 4: Attach seismic cable

For cable assembly, see Cables (page 134).

END OF DETAIL.

In-line Heat Exchanger

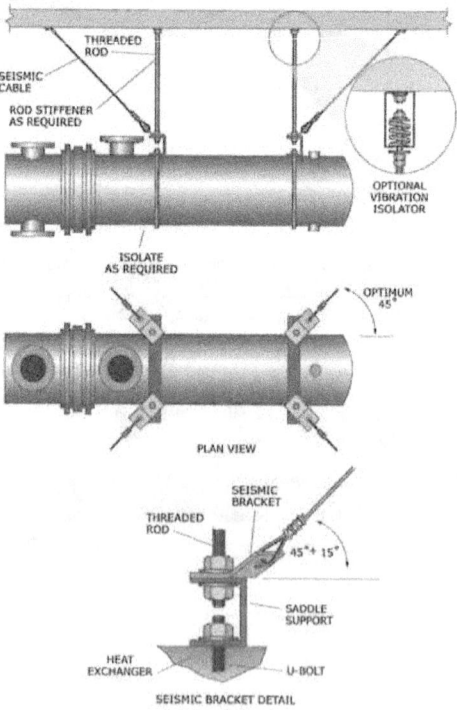

Figure 98: In-line heat exchanger.

Step 1: Attach pipe clamps to the building structure using threaded rods and anchors

Lay out all attachment points before anchoring, then refer to Attachment Details Connecting to Building Structure (page 102). For instructions on installing anchors, see Anchors (page 107).

Step 2: Add rod stiffeners

For rod stiffener details, refer to Figure 86 (page 86).

Step 3: Install anchors for cable bracing

Attach cable seismic brackets with anchors. For typical anchorage to different building construction, refer to Attachment Details Connecting to Building Structure (page 102). For instructions on installing anchors, see Anchors (page 107).

Step 4: Attach seismic cable

For cable assembly, see Cables (page 134).

END OF DETAIL.

ATTACHMENT DETAILS CONNECTING TO BUILDING STRUCTURE

This section details seven types of attachments to building structures:

- To concrete fill on steel deck (this page).
- To wood beam (page 103).
- To I-beam (page 103).
- To bar joist (page 104).
- To concrete slab (page 104).
- Cable brace attachment (page 105).
- Steel shaped brace attachment (page 106).

To concrete fill on steel deck

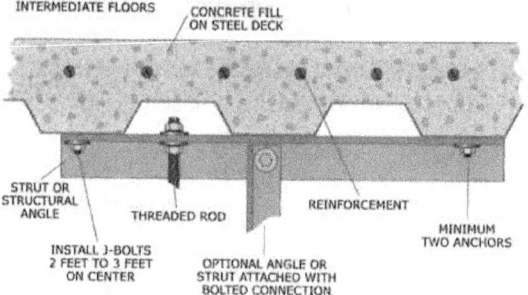

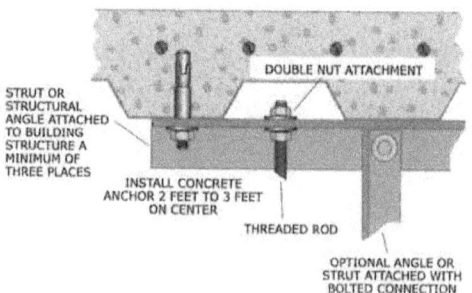

Figure 99: Post-installed anchor; concrete fill on steel deck.

END OF DETAIL.

To wood beam

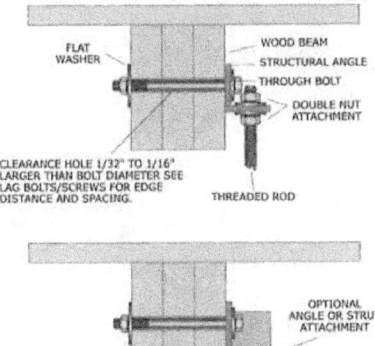

Figure 100: Wood beam construction.

 For edge distance and spacing, see Lag Bolts (page 116).

END OF DETAIL.

To I-beam

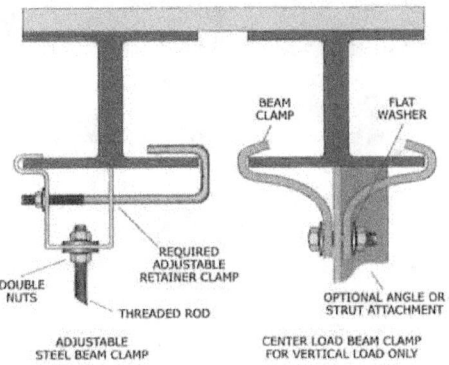

Figure 101: Steel I-beam construction.

 Use center load beam clamps for vertical loads. Do not use for cables, rods, or structural members positioned at an angle.

END OF DETAIL.

Attachment Details Connecting to Building Structure

To bar joist

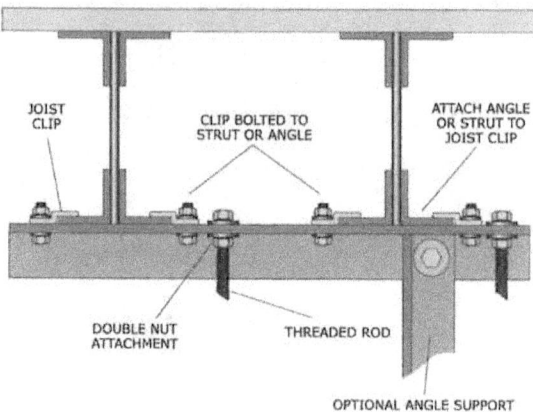

Figure 102: Bar joist construction.

 Use center load beam clamps for vertical loads. Do not use for cables, rods, or structural members positioned at an angle.

END OF DETAIL.

To concrete slab

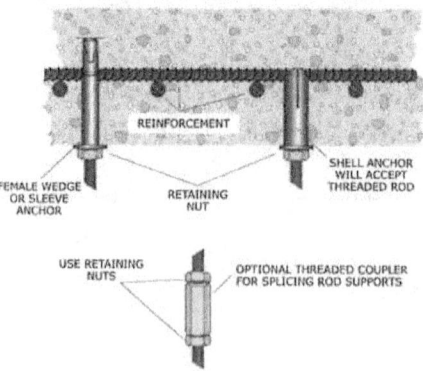

Figure 103: Concrete slab construction.

END OF DETAIL.

Cable brace attachment

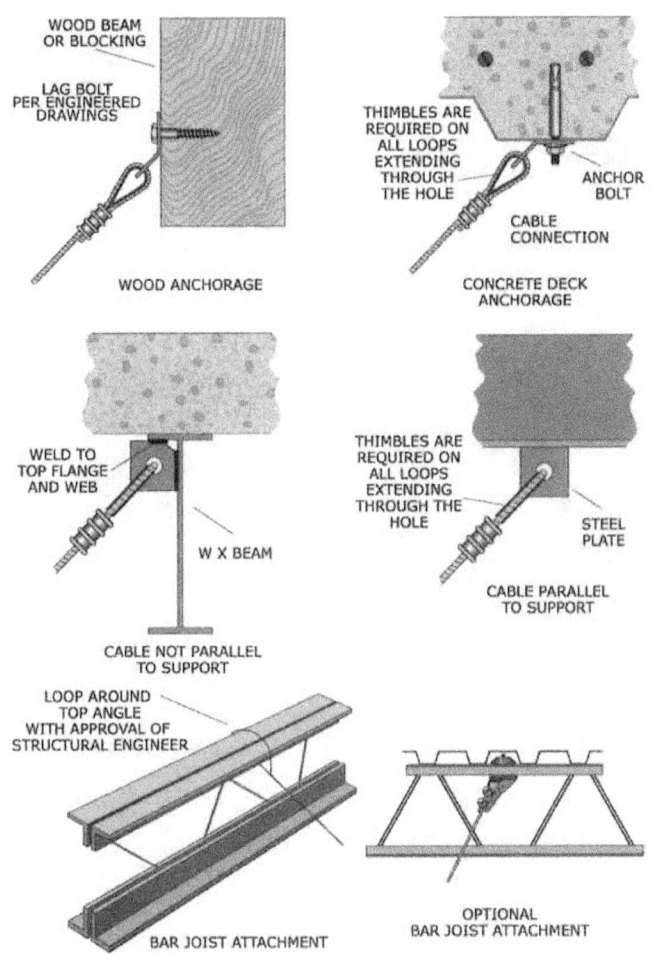

Figure 104: Attachment of cable brace to the building structure.

END OF DETAIL.

Steel shaped brace attachment

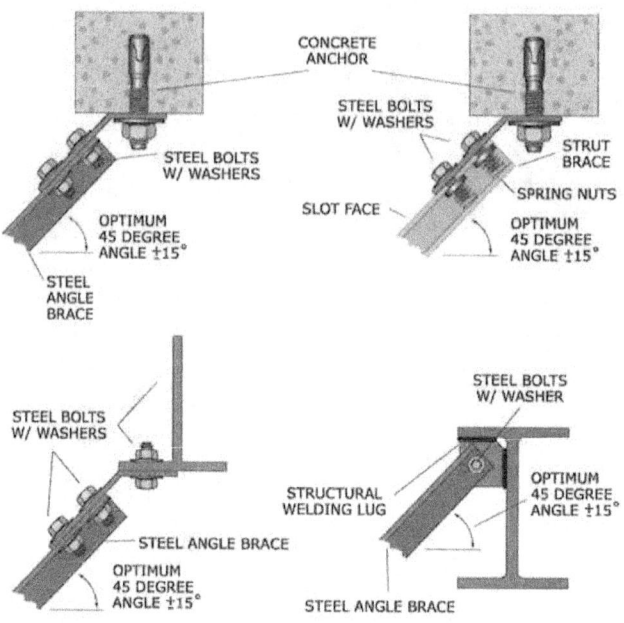

Figure 105: Attachment of angle or strut to the building structure.

END OF DETAIL.

ANCHORS

General Anchors

 IMPORTANT: Installation methods depend on the type of anchor and the particular application. Always follow the anchor manufacturer's installation instructions.

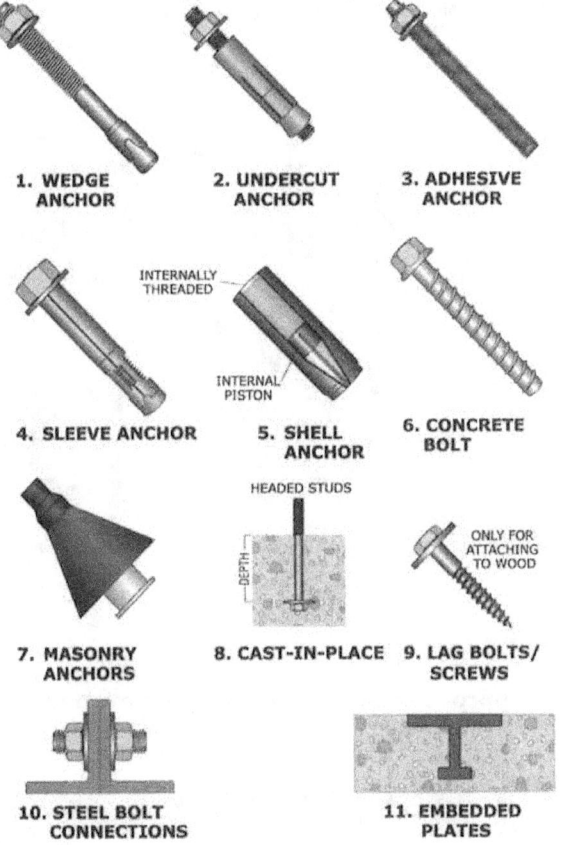

Figure 106: Types of anchors.

 Some anchors must not be used with vibratory loads.

Step 1: Determine the type of anchor

Using Figure 106 (page 107), identify the anchor recommended for your application. Anchors 1-6 are post-installed anchors and instructions for installing them begin on this page. Anchors 7-11 are specialty anchors and instructions are shown on pages 114 to 133.

The various steps for installing anchors into concrete, brick, and concrete block are shown below.

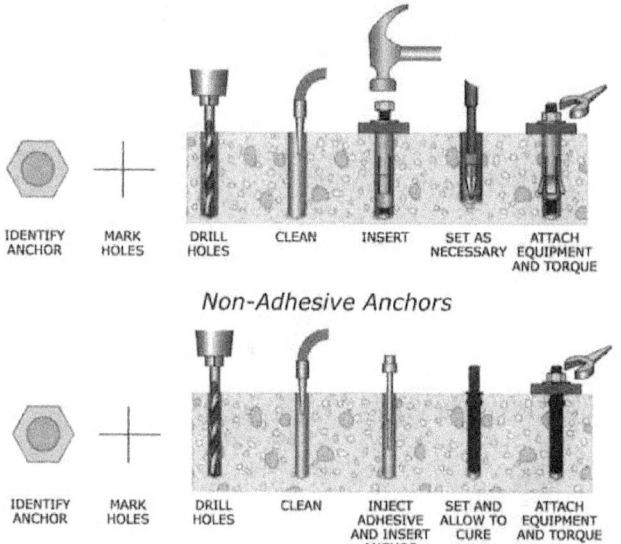

IDENTIFY ANCHOR | MARK HOLES | DRILL HOLES | CLEAN | INSERT | SET AS NECESSARY | ATTACH EQUIPMENT AND TORQUE

Non-Adhesive Anchors

IDENTIFY ANCHOR | MARK HOLES | DRILL HOLES | CLEAN | INJECT ADHESIVE AND INSERT ANCHOR | SET AND ALLOW TO CURE | ATTACH EQUIPMENT AND TORQUE

Adhesive Anchors

Figure 107: Summary of installation steps.

 Approved construction documents may require special inspection to torque anchors or for proof load using hydraulic rams.

Step 2: Determine where to drill the hole

To determine anchor locations for the equipment you are installing, follow the instructions for the equipment, bracing and attachment you are using (pages 16 to 106). Coordinate the equipment connections and hole locations with the location of any steel reinforcements or tendons.

Determine the depth and location of any steel reinforcement or tendons *before* drilling. This may require relocating equipment slightly to avoid the reinforcement.

 FOR POST-TENSIONED (PRE-STRESSED) BUILDINGS, LOCATE THE TENDONS BEFORE DRILLING. EXTREME DAMAGE MAY OCCUR IF A TENDON IS NICKED OR CUT.

When using electronic locating devices to find reinforcement and tendons, make sure you know the limitations of the device. Calibrate and test with a known standard or location to confirm accuracy. Check the area of concern in two directions. Inform the contractor performing the work of the precision of the test unit and record the results. For example: *agreed upon mark +/- ¼" location vertical, horizontal, and depth +/- ½"*.

Coordinate the location of anchors with the edge of the concrete, construction joints, and other anchors.

 Do not install the anchor too close to the edge of the concrete base. Typically, the anchor's distance from the edge is 1½ times the embedment depth.

 Do not install an anchor too close to another anchor. Typically, the minimum spacing between anchors is two times the anchor's embedment depth.

Step 3: Drill the hole

 Drill the right-sized hole for the anchors. Use the appropriate ANSI-rated drill bit for the application.

 Do not drill holes into concrete at an angle.

For wedge, undercut and sleeve anchors, drill the hole deeper than the required embedment depth.

 The required hole depth may be different from the embedment depth. See Figure 108 (page 110).

Anchors: General

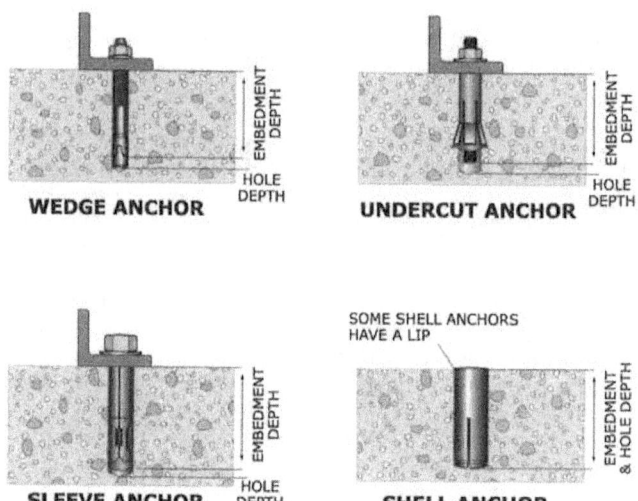

WEDGE ANCHOR

UNDERCUT ANCHOR

SLEEVE ANCHOR

SOME SHELL ANCHORS HAVE A LIP

SHELL ANCHOR

Figure 108: Embedment depth and hole depth of four anchor types.

The depth of the concrete base must be at least one inch greater than the hole you are drilling.

 Some undercut anchors require an even deeper concrete base.

 If you strike steel reinforcement when drilling, you must have the damage inspected. As directed, fill the hole with approved grout and select a new location according to minimum spacing requirements. Drill a new hole (see Figure 109 below).

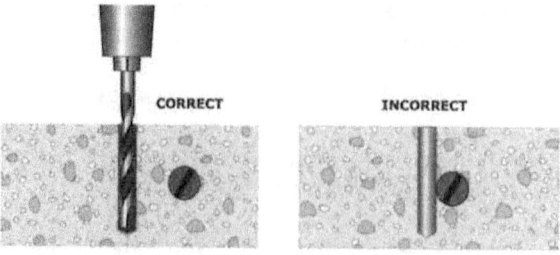

CORRECT

INCORRECT

Figure 109: Drilling into concrete with rebar.

Step 4: Clean out the hole

Drilled holes must be cleaned before you can insert the anchor. Use clean, dry compressed air to blow out dust and debris. The type of anchor or application also may require you to use a brush.

 See the anchor manufacturer's instructions for cleaning the hole.

 CLEANING IS IMPORTANT: a "dirty" hole can significantly reduce an anchor's performance.

Step 5: Insert the anchor

If you are installing any anchor *other than* an adhesive anchor, drive the anchor into the hole with a hammer or use a wrench rotor hammer for concrete bolts.

 IMPORTANT: DO NOT DAMAGE THE THREADS DURING INSTALLATION. DO NOT FORCE THE ANCHOR. If you use a larger hammer than recommended by the manufacturer, you may damage the anchor.

If you are installing an adhesive anchor, insert the capsule or inject non-capsule adhesive into the hole. Slowly rotate the anchor into place as shown in Figure 110 (below).

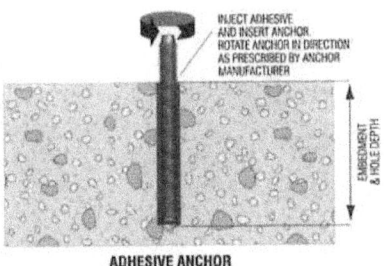

ADHESIVE ANCHOR

Figure 110: Adhesive anchor installation.

If you have installed a wedge or sleeve anchor, go to Step 7 (page 113).

Step 6: Setting adhesive, shell and undercut anchors ONLY

 Adhesive Anchor: Allow enough time for the adhesive to fully cure. The curing process may take a long time. See the manufacturer's instructions.

 Shell Anchor: Drive the prescribed setting tool into the anchor until the setting tool shoulder meets the edge of the anchor, as shown below. See the manufacturer's instructions.

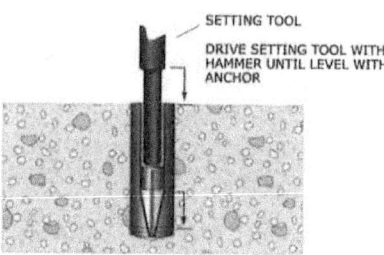

SHELL ANCHOR

Figure 111: Set shell anchors.

 Undercut Anchor: Use special tools provided by the anchor manufacturer to set the anchor, as shown below. See the manufacturer's instructions.

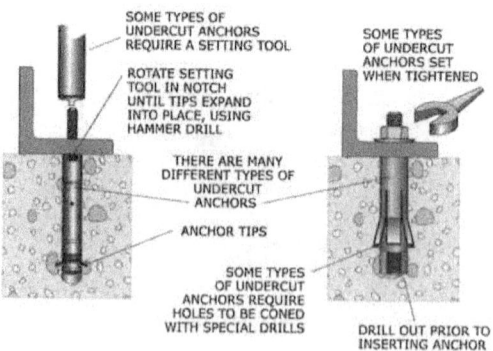

UNDERCUT ANCHORS

Figure 112: Set undercut anchors.

Step 7: Set the equipment and tighten the anchors

Set the equipment in place. Check for gaps. Gaps under the equipment must not be greater than 1/8" as shown below. If the gap is greater than 1/8", dry pack the gap with grout and repeat this step.

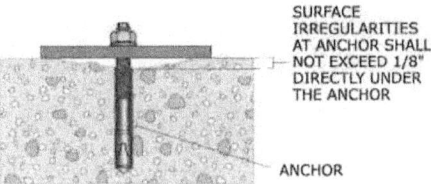

Figure 113: Acceptable gap for grouted plate.

For anchor bolts larger than 3/8", the equipment housing should be reinforced using a structural angle bracket as shown in Figure 114 (below).

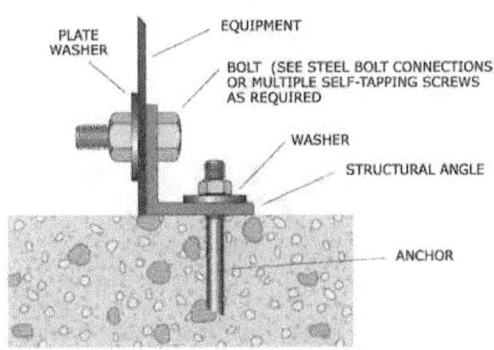

Figure 114: Installing a reinforcing angle bracket to equipment.

 Tighten the anchor bolt to the correct torque setting in the manufacturer's instructions or approved construction documents. Use a calibrated torque wrench.

END OF DETAIL.

113

Cast-in-place Anchors

Cast-in-place anchors are embedded in the concrete when the floors or walls are poured. Bolts are firmly held in place while the concrete is poured to maintain proper alignment and position. The size and location of the anchors can be determined from approved construction documents.

Step 1: Move the equipment into place

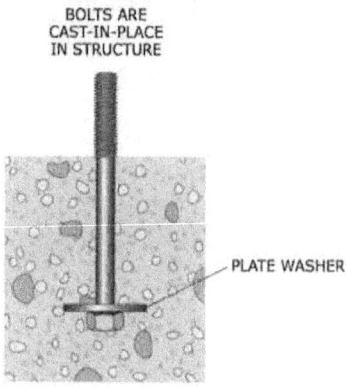

Figure 115: Bolting equipment to cast-in-place anchors.

Step 2: Secure equipment

Once the equipment is in place, apply washers and nuts and then tighten.

 Tighten the anchor bolt to the correct torque setting in the manufacturer's instructions or approved construction documents.

Use a calibrated torque wrench or turn-of-nut method (see Table 5, page 127).

END OF DETAIL.

Lag Bolts

Lag bolts are used to attach equipment or steel shapes to wood structures. The size and location of the anchors can be determined from approved construction documents (see Figure 116, below).

- The edge distance for the non-loaded side is 1½ times the bolt diameter.

- The edge distance for the loaded side is 4 times the bolt diameter.

- The spacing between bolts is 4 times the bolt diameter.

- The end distance is 7 times the bolt diameter.

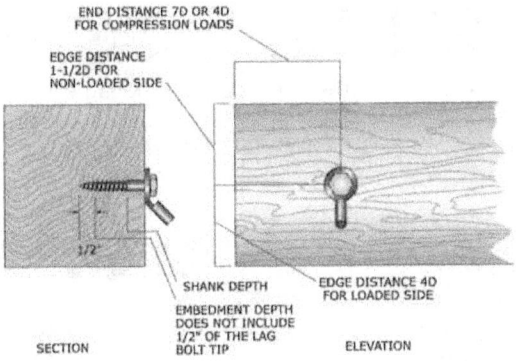

Figure 116: Spacing requirements for wood lag bolts.

 Figure 116 is allowed for cable attachments. If a strut is attached to wood, both edge distances are considered loaded and require 4 times the bolt diameter.

 Do not anchor to the end grain.

Step 1: Mark the location of the lag bolts

Lead holes and clearance holes are not required for lag bolts that are 3/8" or smaller. If the lag bolt is smaller than 3/8", go to Step 4 (page 116).

Step 2: Drill a clearance hole

Drill a hole with a drill bit the same size as the shank of the bolt. The depth of the hole is the same as the length of the unthreaded shank that will extend into the wood (see Figure 116, page 115).

Step 3: Drill a lead hole

Drill a hole with a drill bit that is 60% to 70% of the diameter of the shank of the bolt. The depth of the hole is the same as the embedment depth of the bolt.

Step 4: Move the equipment or steel shape into place

Step 5: Install the lag bolt with a wrench

You may use soap or other lubricant on the lag bolt.

 DO NOT USE A HAMMER TO INSTALL LAG BOLTS.

Step 6: Tighten the bolt

Hand-adjust the lag bolt until there is firm contact between the lag bolt and connected metal components. Hand tools may be used to bring the lag bolt and metal components into contact. Following contact, tighten nut as shown in Table 5 (page 127).

END OF DETAIL.

Masonry and Drywall Anchors

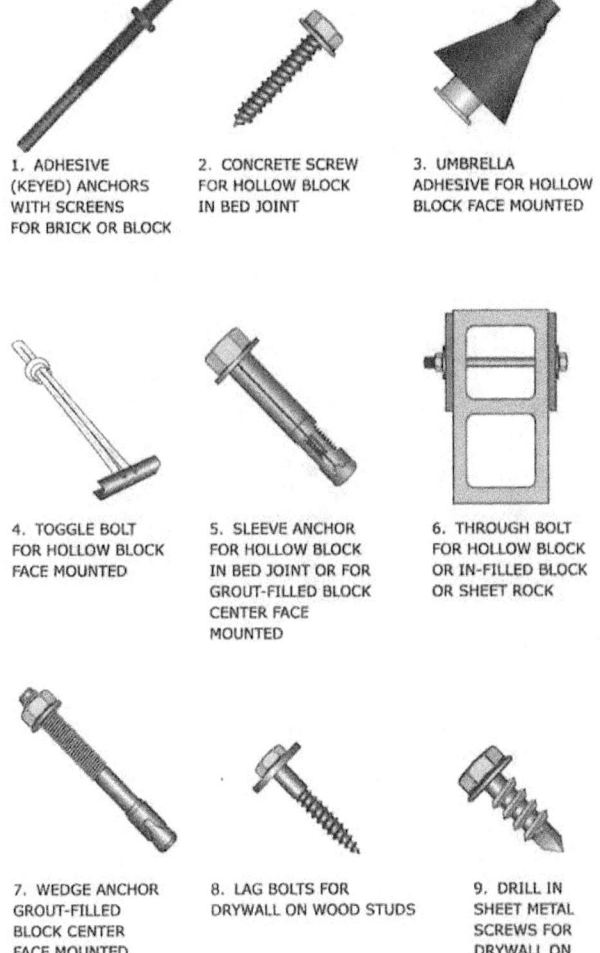

1. ADHESIVE (KEYED) ANCHORS WITH SCREENS FOR BRICK OR BLOCK

2. CONCRETE SCREW FOR HOLLOW BLOCK IN BED JOINT

3. UMBRELLA ADHESIVE FOR HOLLOW BLOCK FACE MOUNTED

4. TOGGLE BOLT FOR HOLLOW BLOCK FACE MOUNTED

5. SLEEVE ANCHOR FOR HOLLOW BLOCK IN BED JOINT OR FOR GROUT-FILLED BLOCK CENTER FACE MOUNTED

6. THROUGH BOLT FOR HOLLOW BLOCK OR IN-FILLED BLOCK OR SHEET ROCK

7. WEDGE ANCHOR GROUT-FILLED BLOCK CENTER FACE MOUNTED

8. LAG BOLTS FOR DRYWALL ON WOOD STUDS

9. DRILL IN SHEET METAL SCREWS FOR DRYWALL ON METAL STUDS

Figure 117: Types of masonry and drywall anchors.

Step 2: Determine where to drill the hole

Anchors shown in Figure 117 (page 117) must be installed in specific areas of hollow block and in-filled block. See Figure 118 (below) for approved anchor hole locations when using any of the concrete block anchors shown in Figure 117.

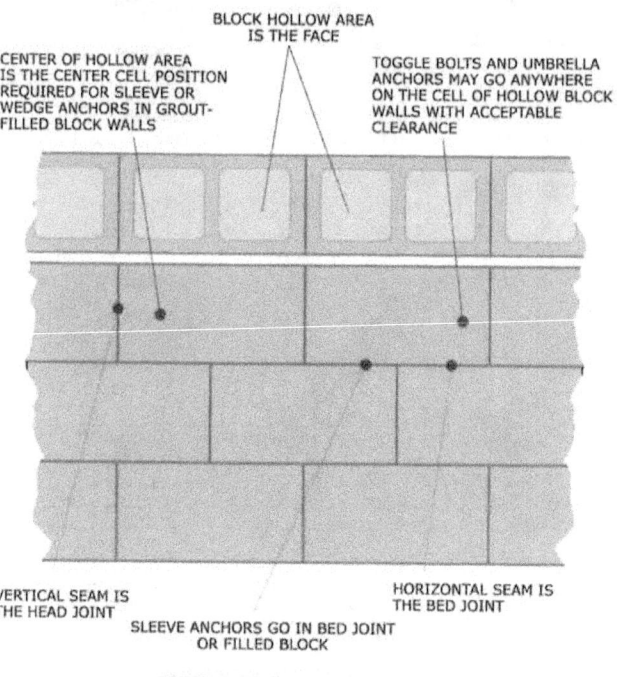

CENTER OF HOLLOW AREA
IS THE CENTER CELL POSITION
REQUIRED FOR SLEEVE OR
WEDGE ANCHORS IN GROUT-
FILLED BLOCK WALLS

BLOCK HOLLOW AREA
IS THE FACE

TOGGLE BOLTS AND UMBRELLA
ANCHORS MAY GO ANYWHERE
ON THE CELL OF HOLLOW BLOCK
WALLS WITH ACCEPTABLE
CLEARANCE

VERTICAL SEAM IS
THE HEAD JOINT

HORIZONTAL SEAM IS
THE BED JOINT

SLEEVE ANCHORS GO IN BED JOINT
OR FILLED BLOCK

CONCRETE SCREWS CAN GO ANYWHERE

Figure 118: Block wall locations.

The location of the anchors should be coordinated with the block webs, or centered in the cell face, *and properly spaced from other anchors.*

 DO NOT POSITION THE HOLES IN THE HEAD JOINT. Carefully note the location of anchors in the face location, centered face location, and bed joint as they apply to different anchors.

Find webs in block walls, reinforcements in grout-filled walls, or studs in drywall. Location of through-bolts: Avoid webs in concrete block and studs in drywall walls.

 Location of lag bolts and sheet metal screws in drywall: Install in a wood or metal stud as appropriate.

 Determine the depth and location of any steel reinforcement in grout-filled block walls or brick before drilling. This may require relocating equipment slightly to miss reinforcement.

When using electronic locating devices to find reinforcements and tendons, make sure you know the limitations of the device. Calibrate and test with a known standard or location to confirm accuracy. Check the area of concern in two directions. Inform the contractor performing the work of the precision of the test unit and record the results. For example: *agreed upon mark +/- ¼" location vertical, horizontal, and depth +/- ½"*.

Step 3: Drill the hole

 Drill the right-sized hole for the anchors. Use the appropriate ANSI-rated drill bit for the application.

Use masonry drill bits for brick and block.

 DO NOT CUT STEEL REINFORCEMENT WHEN DRILLING HOLES.

 If you strike steel reinforcement when drilling, you must have the damage inspected. As directed, fill the hole with approved grout and select a new location according to minimum spacing requirements. Drill a new hole (see Figure 109, page 110).

 Holes for concrete screws are smaller than screw size. See the manufacturer's instructions for specific requirements.

Step 4: Clean out the hole

Drilled holes must be cleaned before you can insert the anchor. Use clean, dry compressed air to blow out dust and debris. The type of anchor or application also may require you to use a brush.

 See the anchor manufacturer's instructions for cleaning the hole.

 CLEANING IS IMPORTANT: a "dirty" hole can significantly reduce an anchor's performance.

Step 5: Insert the anchor

The following anchors use different insertion methods:

- Adhesive screen anchor in a brick wall or hollow block wall (this page).
- Adhesive anchor in a hollow block wall (page 121).
- Concrete screw (page 122).
- Toggle bolt (page 122).
- Concrete anchor (sleeve anchor or wedge anchor) (page 123).
- Drywall anchor (lag bolts and sheet metal screws) (page 123).

Adhesive screen anchor in a brick wall or hollow block wall

 See the anchor manufacturer's instructions before connecting the anchor to a brick or hollow block wall.

A screen insert is shown in Figure 119 (page 121). Insert the screen in the wall. Inject the adhesive. Slowly insert the anchor with a twisting motion.

 Screens may be filled with adhesive before inserting the screen into the hole.

For details on installing adhesive anchors in a brick wall, see Figure 120 (below). Similar installation applies to hollow block walls. Adjust the anchor by hand while the adhesive sets.

 DO NOT TOUCH THE ANCHOR WHILE THE ADHESIVE IS CURING.

METAL SCREEN TUBES

Figure 119: Brick/block wall insert.

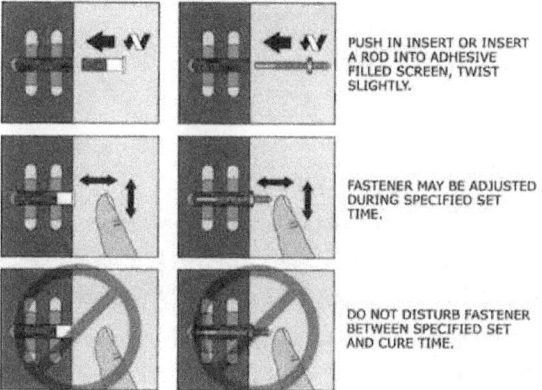

PUSH IN INSERT OR INSERT A ROD INTO ADHESIVE FILLED SCREEN, TWIST SLIGHTLY.

FASTENER MAY BE ADJUSTED DURING SPECIFIED SET TIME.

DO NOT DISTURB FASTENER BETWEEN SPECIFIED SET AND CURE TIME.

Figure 120: Brick wall adhesive anchor.

Adhesive anchor in a hollow block wall

 See the anchor manufacturer's instructions before connecting the anchor to a hollow block wall.

Push an umbrella anchor into the hole until the umbrella unfolds in the block cavity (Figure 121, page 122). Inject adhesive into the umbrella. Slowly insert stud or fastener with a twisting motion.

 DO NOT LEAK ADHESIVE ON THE THREADED PORTION OR CLEAN WITH SOLVENT. The threaded area must be free of debris to attach to a threaded rod or steel bolt.

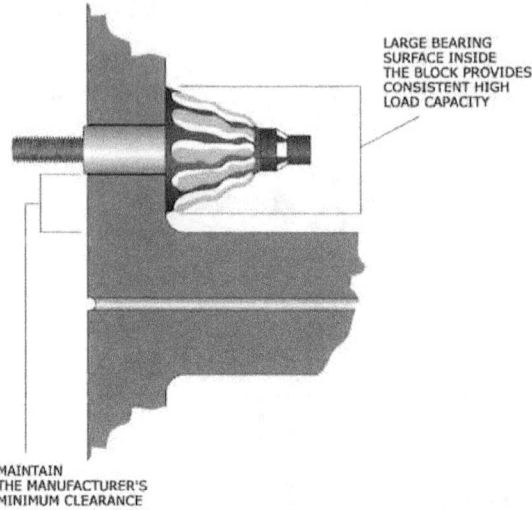

UMBRELLA INSERTS – SPECIFICALLY DESIGNED FOR FASTENING
TO THE FACE OF CONCRETE BLOCK, CLAY TILE OR TERRA COTTA.
ACCEPTS RODS BETWEEN 1/4" AND 1/2"

LARGE BEARING
SURFACE INSIDE
THE BLOCK PROVIDES
CONSISTENT HIGH
LOAD CAPACITY

MAINTAIN
THE MANUFACTURER'S
MINIMUM CLEARANCE

Figure 121: Umbrella anchor in a hollow block wall.

Concrete screw

Drill bits may be specifically sized for each manufacturer,
and typically are smaller in diameter than the nominal or
fractional diameter of a screw. Install a concrete screw with
a rotary drill and bolt the head attachment.

Toggle bolt

Hold the toggle flat alongside the plastic straps and slide the
channel through the hole. Slide the holding ring toward the
wall until the channel is flush with the wall. Cut off the
straps at the holding ring. Insert the bolt with a rotary drill
over the bracket or equipment mounting. See Figure 122
(page 123).

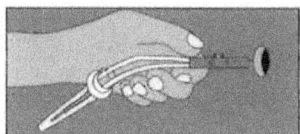

DRILL HOLE. HOLD TOGGLE FLAT
ALONGSIDE PLASTIC STRAPS, SLIDE
THROUGH HOLE.

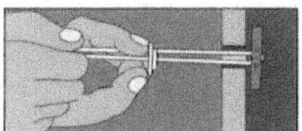

WITH ONE HAND, PULL RING SO METAL
CHANNEL RESTS FLUSH BEHIND WALL.
SLIDE PLASTIC CAP ALONG STRAPS WITH
OTHER HAND UNTIL FLANGE OF CAP IS
FLUSH WITH WALL.

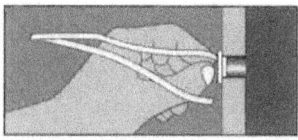

PLACE THUMB BETWEEN PLASTIC STRAPS.
PUSH SIDE TO SIDE SNAPPING OFF
STRAPS FLUSH WITH WALL.

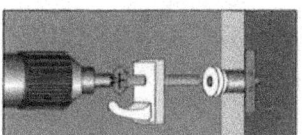

INSERT BOLT THROUGH ITEM TO BE
ATTACHED AND TIGHTEN UNTIL FLUSH
WITH FIXTURE. MINIMUM CLEARANCE
BEHIND WALL: 1-7/8" (48MM).

Figure 122: Toggle bolt installation.

 DO NOT OVER-TIGHTEN.

Concrete anchor (sleeve anchor or wedge anchor)

Use a hammer to drive the anchor in the hole.

 **DO NOT FORCE THE ANCHOR. If you use a
hammer larger than recommended, you may
damage the anchor.**

To determine the embedment depth of post-installed
anchors, see Figure 108 (page 110).

Drywall anchor (lag bolts and sheet metal screws)

Use a rotary drill to insert the anchor.

 DO NOT OVER-TIGHTEN.

Step 6: Set the anchor (adhesive only)

Allow enough time for the adhesive to harden and adhere to the concrete. *This may take several hours.*

Step 7: Set the equipment and tighten the anchors

 Tighten the anchor bolt to the proper torque setting as shown in the anchor manufacturer's instructions or approved construction documents.

In-filled block walls may have gaps in the grout fill or the grout may slightly crack, requiring anchors to be installed in the center of the cell.

 If the grout cracks severely, or if you miss a grouted block, the anchor will not tighten and will pull out. If it pulls out, move the anchor to a new centered cell location.

END OF DETAIL.

Power-Actuated Anchors

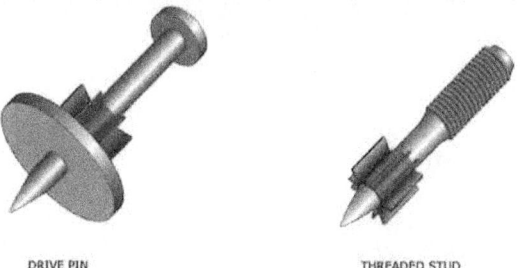

DRIVE PIN   THREADED STUD

Figure 123: Two types of power-actuated anchors.

> (!) **Authorities having jurisdiction may require a
>   license to use power-actuated anchors.**

> (!) **Refer to the Anchor Selection Guide (page 148)
>   for additional information.**

 **Follow the manufacturer's recommendations
for minimum slab thickness and best location.**

Using tools provided by the anchor manufacturer, install the
anchor at the desired location. Follow all the manufacturer's
instructions. Also see Power-Actuated information in the
Anchor Selection Guide (page 148).

For threaded rods, follow Step 4.

When using drive pins, install the pin with the strap or wire
to complete the installation.

Tighten the anchor nuts over equipment brackets on the
threaded rods.

END OF DETAIL.

Steel Bolt Connections

The three ways to attach bolted connections are:

- Connecting the base of the equipment to an angle bolted to a concrete floor (this page).
- Bolting two structural steel shapes together (page 128).
- Bolting a threaded rod to steel shapes or strut (page 129).

Connecting the base of the equipment to an angle bolted to a concrete floor

Step 1: Preparation

 Determine the bolt size or sheet metal screw and material requirements from approved construction documents or the manufacturer's instructions.

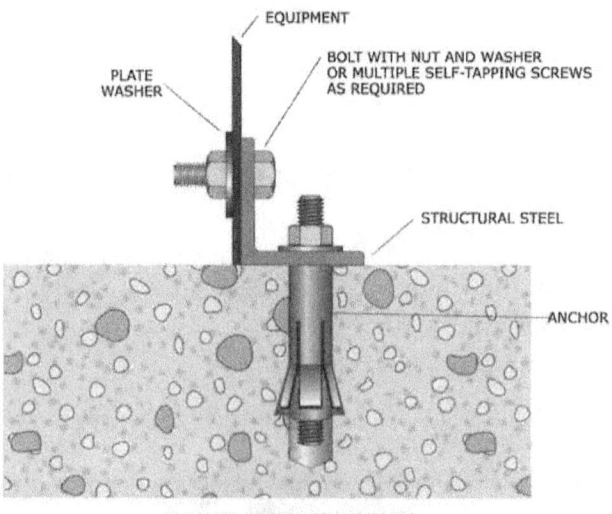

EQUIPMENT

BOLT WITH NUT AND WASHER
OR MULTIPLE SELF-TAPPING SCREWS
AS REQUIRED

PLATE
WASHER

STRUCTURAL STEEL

ANCHOR

USE PLATE WASHER TO REINFORCE
LIGHT SHEET METAL HOUSINGS

Figure 124: Bolting equipment to an angle.

Step 2: Locate holes

Use pre-drilled holes wherever possible. Holes may not have been pre-drilled at the attachment locations shown in the instructions. In these cases, carefully drill new holes in the correct locations.

 Use caution when drilling into equipment. Internal components can be damaged. DO NOT DRILL OVERSIZED HOLES.

Step 3: Install bolts, washers, and nuts

Once the equipment is in place, apply washers and nuts and then tighten.

 Tighten the anchor bolt to the correct torque setting shown in the manufacturer's instructions or approved construction documents.

For turn-of-nut tightening hand-adjust the bolt snug tight where there is firm contact between the bolt and connected metal components. Hand tools may be used to bring the bolt an meta components into contact. Following contact, tighten the nut as shown below.

Length of Bolt	Additional Tightening
Up to and including 4 diameters	1/3 turn
Over 4 diameters and not more than 8 diameters	1/2 turn
Over 8 diameters and not more than 12 diameters	2/3 turn

Table 5: Turn-of-nut, hand-adjusted tightening.

Bolting two structural steel shapes together

Step 1: Preparation

 Determine the bolt size and material requirements from approved construction documents or the manufacturer's instructions.

Figure 125: Bolting structural shapes.

Step 2: Locate holes

Carefully drill new holes in the structural steel shapes.

Step 3: Install bolts, washers, and nuts

Apply washers and nuts, then tighten.

 Tighten the anchor bolt to the correct torque setting shown in the manufacturer's instructions or approved construction documents. Use a calibrated torque wrench or turn-of-nut method (see Table 5, page 127).

END OF DETAIL.

Bolting a threaded rod to steel shapes or strut

A threaded rod is used with suspended equipment. This section includes attachment to the equipment and attachment to the building structure (page 102).

Step 1: Preparation

 Determine the threaded rod size from approved construction documents or the manufacturer's instructions.

The three different ways to attach the threaded rod are shown in Figure 126 (below).

Step 2: Attach the top connection of the threaded rod

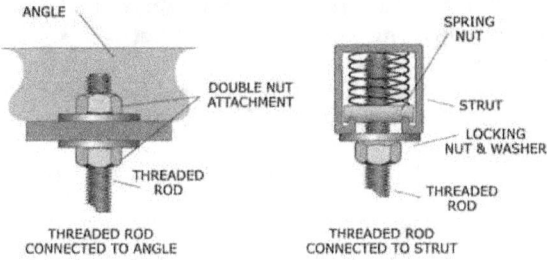

THREADED ROD
CONNECTED TO ANGLE

THREADED ROD
CONNECTED TO STRUT

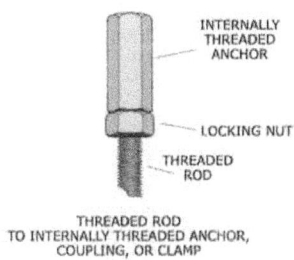

THREADED ROD
TO INTERNALLY THREADED ANCHOR,
COUPLING, OR CLAMP

Figure 126: Attaching the top connection of the threaded rod.

Apply washers and nuts, then tighten.

Tighten the anchor bolt to the correct torque setting shown in the manufacturer's instructions or approved construction documents. Use a calibrated torque wrench or turn-of-nut method (see Table 5, page 127).

Step 3: Attach threaded rods to equipment brackets

Equipment without attachment brackets requires additional steel shapes for connections to the building structure and/or roof.

Once the equipment is in place, apply washers and nuts, then tighten.

Tighten the anchor bolt to the correct torque setting shown in the manufacturer's instructions or approved construction documents. Use a calibrated torque wrench or turn-of-nut method (see Table 5, page 127).

END OF DETAIL.

Welding

 Before welding, refer to approved construction documents and specifications, seismic restraint submittals, and manufacturer's instructions.

Attaching equipment to embedded plates: Plates are embedded in the concrete during the floor or wall pour. Plates are firmly held in place while the concrete is poured to maintain proper alignment and position. The size and location of the plate can be determined from construction drawings. See Figure 127 (below) for weld locations.

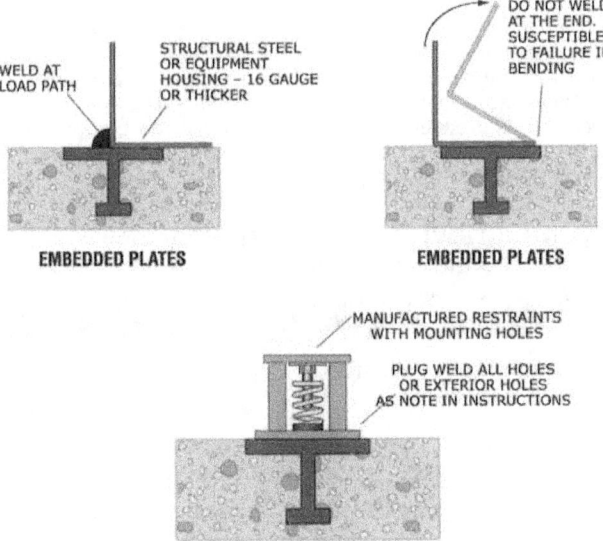

Figure 127: Welding to embedded plates.

Attaching structural shapes and plates: Shapes and plates are welded to provide equipment attachment. All weld base material must be thick enough for the weld size specified.

Step 1: Determine the weld material, shape, and dimensions for each piece

Step 2: Fit the material to ensure proper weld joint preparation

Step 3: Clean the surfaces

Surfaces must be dry and free of galvanized coating, hot-dipped or rust inhibitor, paint, scale, rust, oil, grease, water, and other foreign material for a minimum of one inch from the estimated toe of the weld.

Step 4: Weld the materials

The weld must be as prescribed in the welding procedure specifications (WPS).

> WPS for shop and field pre-qualified weld joints and weld joints qualified by test must be prepared for review and approval before fabrication. All welding procedure items such as base metals, welding processes, filler metals and joint details that meet the requirements of AWS D1.1 will be considered prequalified. Any change or substitution beyond the range or tolerance or requirements for pre-qualification will be qualified by test pre-AWS D1.1.

 DO NOT WELD OVER PAINT. You may paint after welding has cooled to room temperature.

Step 5: Inspect the weld

Make sure the surface is free of slag, dirt, grease, oil, scale, or other contaminants.

Welds cannot have cracks. Adjacent layers of weld metal and base metal must be thoroughly fused together.

All craters must be filled to the full cross-section except outside the effective weld length.

Underrun must not exceed 1/16". Undercut must not exceed 1/16" for any 2" per 12" weld or 1/32" for the entire weld.

Surfaces must be free of coarse ripples, grooves, abrupt ridges, and valleys. The faces of fillet welds must be flat or slightly convex.

END OF DETAIL.

Anchor Sizes for Suspended Equipment

Rigidly attach equipment to building structure rods and cables according to Table 6 (below). Torque anchors according to the manufacturer's instructions.

Vertical threaded rod	Quantity	Anchors per Rod
1/2" rod with rod stiffener	One on each corner	1

Rod anchor	Embedment (in.)	Minimum Edge Distance
1/2" wedge	2-1/4"	3-1/2"
1/2" sleeve	2-1/4"	3-1/2"
3/8" lag bolt to wood beam	2"	2" to edge of wood and 3" from end
1/2" steel bolt	N/A	1"

Cable	Quantity	Anchors per cable
3/16" seismic cable	4 at 45 degrees at each corner	1

Cable anchor	Embedment (in.)	Minimum Edge Distance (in.)
1/2" wedge	2-1/4"	3-1/2"
1/2" sleeve	2-1/4"	3-1/2"
3/8" lag bolt to wood beam	2"	2" to edge of wood and 3" from end
1/2" steel bolt	N/A	1"

Table 6: Suspended equipment anchor sizes for rigid connections.

SPECIAL CASES

Cables

The three ways to assemble cable connections are by using:

- Bolts with center holes (page 135).
- Ferrule clamps (page 136).
- Wire rope grips (page 138).

Other end fittings may be acceptable.

Cables should be installed at a 45-degree slope. Where interferences are present, the slope may be a minimum of 30 degrees and a maximum of 60 degrees.

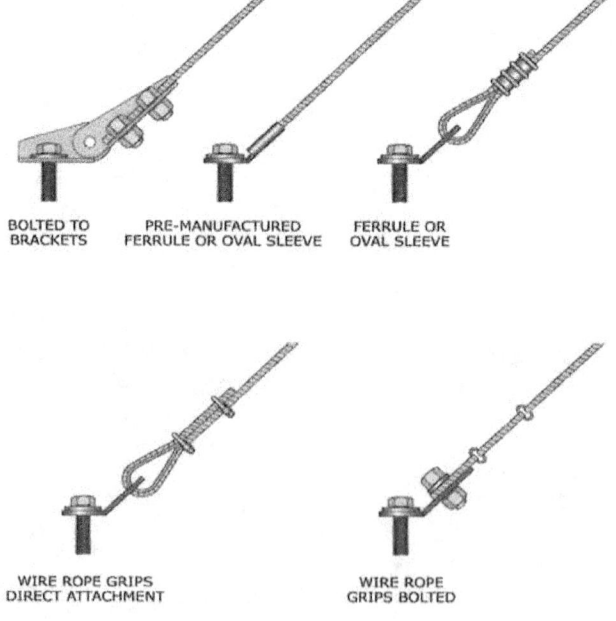

BOLTED TO BRACKETS   PRE-MANUFACTURED FERRULE OR OVAL SLEEVE   FERRULE OR OVAL SLEEVE

WIRE ROPE GRIPS DIRECT ATTACHMENT   WIRE ROPE GRIPS BOLTED

Figure 128: Cable attachments.

Bolts with center holes

The manufacturer provides this type of cable assembly, along with the cables, mounting bolts with holes, and brackets that attach directly to the building structure or equipment frame. Assemble the cable as shown below.

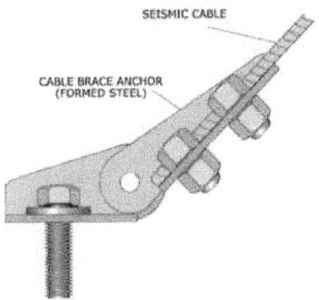

Figure 129: Cable attached with bolts to a bracket.

Step 1: Drill anchor holes in the building structure as required

Step 2: Attach brackets to both the building and the equipment frame

Step 3: Cut the cable to desired length and slide it through the holes in the bolts

Step 4: Tighten the cable

For rigid connections, pull the cable hand tight. Pull the cable hand-tight and let out 1/8" slack for vibration-isolated components. Avoid using too much tension or too much slack.

Step 5: Torque bolts

 Refer to the manufacturer's instructions.

 Overtorque of bolts may cause damage to cables.

END OF DETAIL.

Ferrule clamps

Ferrule clamps may be connected to various types of attachments. Figure 130 (below) and Figure 131 (page 137) show attachments and identify the parts, ferrules or sleeves and thimbles, used in the assembly.

 Ferrules must be made of steel, zinc-plated copper, or steel alloys (including stainless steel). Do not use aluminum ferrules.

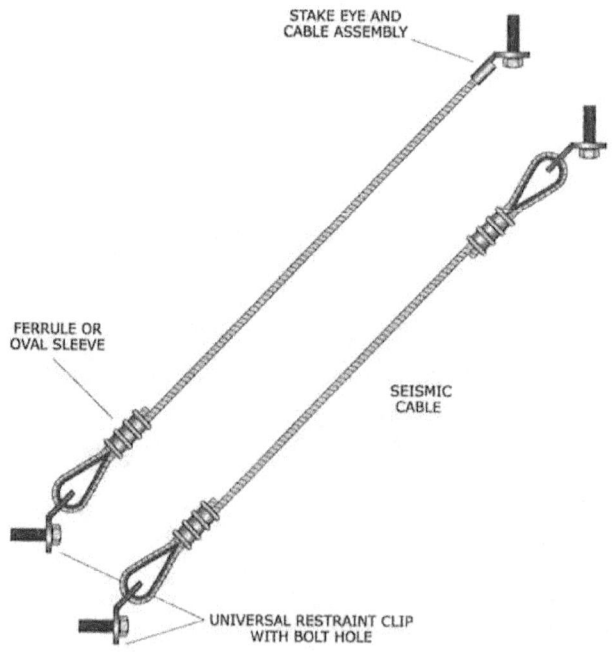

STAKE EYE AND
CABLE ASSEMBLY

FERRULE OR
OVAL SLEEVE

SEISMIC
CABLE

UNIVERSAL RESTRAINT CLIP
WITH BOLT HOLE

Figure 130: Ferrule assemblies.

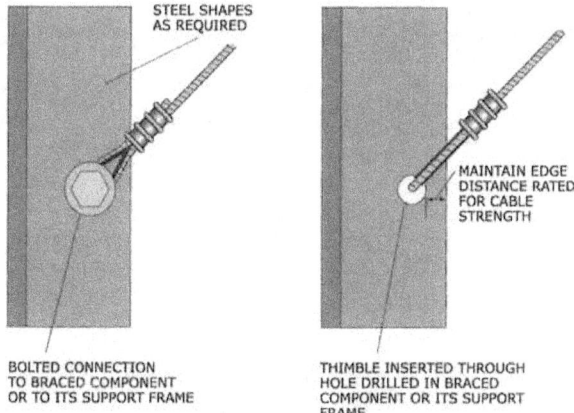

Figure 131: Ferrule attachments.

Step 1: Install brackets with mounting holes, eye-bolts, or drill mounting holes

Install brackets with mounting holes to the structure. Attach cables to the top of cord angles. See Attachment Details Connecting to Building Structure (page 102).

Step 2: Cut the cable to the desired length and slide the oval ferrule (sleeve) onto the cable

Step 3: Wrap the cable around the thimble and pass it through the mounting bolt or holes and back through the ferrule

Step 4: Tighten the cable

For rigid connections, pull the cable tight. For isolated components, leave a small amount of slack. Avoid using too much tension or too much slack.

Step 5: Crimp the ferrule or oval sleeve two or three times as specified in the cable or ferrule manufacturer's instructions

Use crimp tools and gauges specified by the manufacturer. Crimp and verify the depth of the crimp using a gauge.

END OF DETAIL.

137

Wire rope grips

Installing cables attached with wire rope grips is similar to attaching ferrule clamps, as shown below.

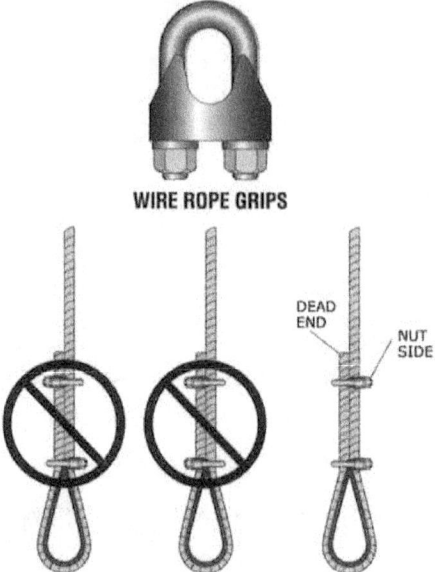

WIRE ROPE GRIPS

Figure 132: Wire rope grip assemblies.

 Note that the nut side must be located opposite the dead end of the cable.

Step 1: Install brackets with mounting holes, eye-bolts, or drill mounting holes

Step 2: Cut cable to the desired length and slide three wire rope grips and thimbles onto the cable

Step 3: Pass the cable through the mounting bolt or holes provided and then back through each of the wire rope grips

 Use thimbles for all cable installations with wire rope grips.

Step 4: Tighten the cable

For rigid connections, pull the cable tight. For isolated
components, leave a small amount of slack. Avoid using too
much tension or too much slack.

Step 5: Torque all bolts evenly

Use the turn-of-nut tightening method described in Step 3
of Steel Bolt Connections (page 126).

 DO NOT OVER-TIGHTEN.

END OF DETAIL.

Flexible Connections and Expansion Joints

Flexible connections are required when vibration-isolated equipment is attached to rigidly supported piping systems or ductwork.

Flexible connections are also required when rigidly mounted equipment is attached to unrestrained piping systems or to unrestrained ductwork systems.

Expansion joints are required when piping systems or ductwork span two different building structures or when entering an isolated building.

Flexible Connectors

The four flexible connectors are:

- Rubber joint duct connector (this page).
- Braided hose pipe connector (page 141).
- Rubber hose pipe connector (page 141).
- Rubber hose connector with control rods (page 141).

 Follow manufacturer's instructions to install connectors.

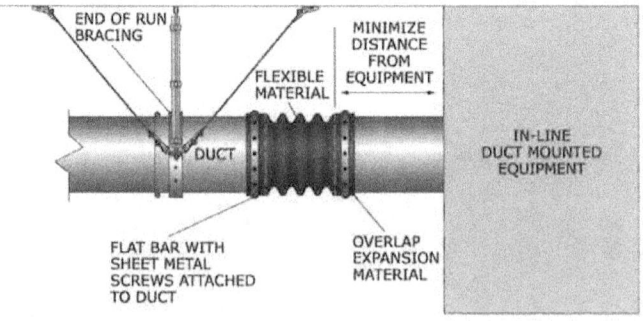

Figure 133: Rubber joint duct connector.

 Rectangular ducts may use slip joints.

 End-of-run seismic braces are required at flexible connections.

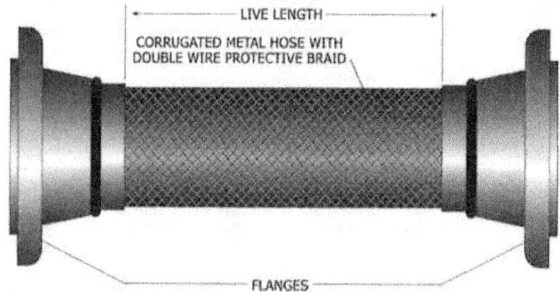

Figure 134: Braided hose pipe connector.

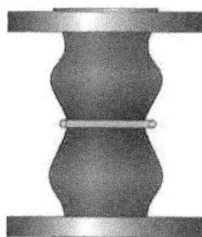

Figure 135: Rubber hose pipe connector.

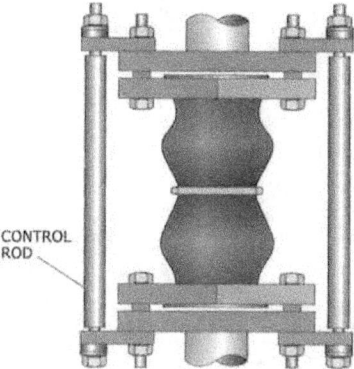

Figure 136: Rubber hose connector with control rods.

END OF DETAIL.

Expansion Joints

The four expansion joints are:

- Braided hose expansion joint (this page).
- Double-socket expansion joint (this page).
- Expansion joint above floor (page 143).
- Expansion joint between a separation (page 143).

Step 1: Verify clearance allowed by expansion joint with movement of the two systems

 Layout all attachment points before making final connections.

 End-of-run seismic braces are required at expansion joints.

Step 2: Run pipe as required by approved construction documents

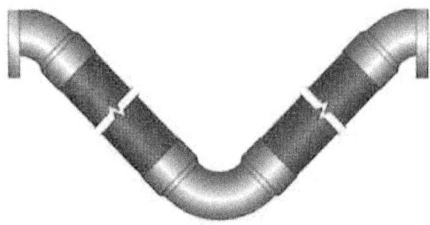

Figure 137: Braided hose expansion joint.

Figure 138: Double-socket expansion joint.

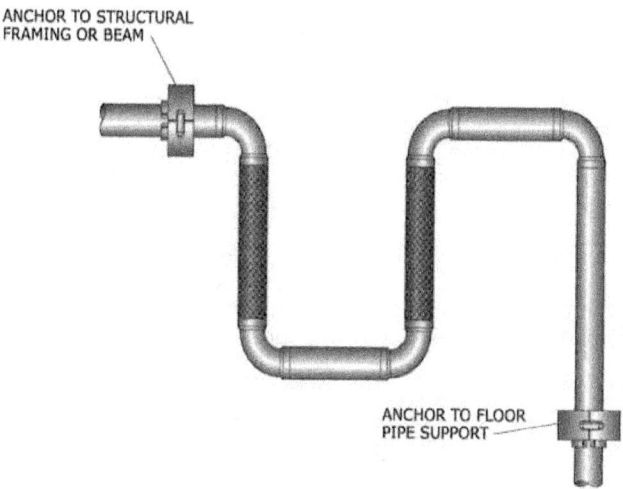

Figure 139: Expansion joint above floor.

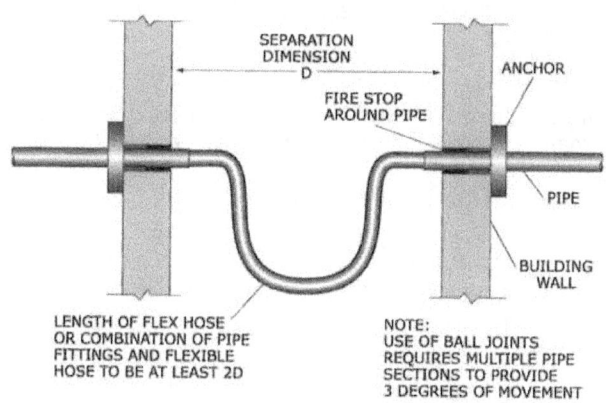

Figure 140: Expansion joint between a separation.

END OF DETAIL.

Valves and Valve Actuators

There are many ways to brace valves and valve actuators. Valves usually do not require special braces and usually are required only in the horizonal direction. Approved construction documents must identify all directions of bracing required.

Rigid or flexible pipe braces near the valve are common. These braces should be installed for specifically bracing the valve and not for bracing the piping system. The piping system should have an independent bracing system.

See Pipe Bracing Details and Installation Instructions (page 54) for pipe braces that can be used near the valve bodies.

There are two types of valve bracing:

- Pipe bracing (page 145).
- Valve bracing (page 146).

Valve actuator seismic braces must be identified on the approved construction documents. These braces are typically unique and depend on the installation and type of valve actuator.

Valve actuators that are directly on top of the valve only require horizontal braces. Heavy actuators located on the side may require additional vertical braces, as detailed in:

- Valve actuator bracing (page 147).

Pipe Bracing

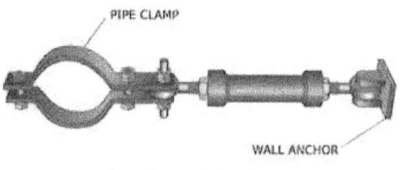

PIPE CLAMP

WALL ANCHOR

HORIZONTAL SUPPORT BRACE

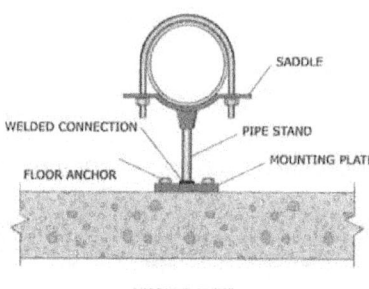

SADDLE

WELDED CONNECTION

PIPE STAND

MOUNTING PLATE

FLOOR ANCHOR

VERTICAL BRACE

Figure 141: Pipe bracing detail.

Step 1: Run piping as shown on the approved construction documents

 Lay out all attachment points before making final piping connections.

Step 2: Verify lengths used for the pipe bracing

Temporarily fit bracing as necessary.

Step 3: Attach mounting plates to the building structure

For instructions on installing anchors, see Anchors (page 107).

Step 4: Assemble brace

END OF DETAIL.

Valve Bracing

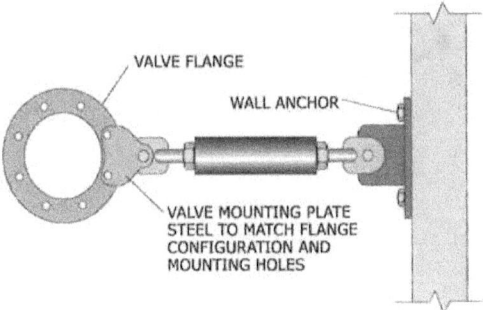

Figure 142: Transverse bracing directly attached to valve.

Step 1: Run piping as shown on the approved construction documents

 Lay out all attachment points before making final piping connections.

Step 2: Verify lengths used for the pipe bracing

Temporarily fit bracing as necessary.

Step 3: Attach mounting plates to the building structure

For instructions on installing anchors, see Anchors (page 107).

Step 4: Assemble bracing

END OF DETAIL.

Valve Actuator Bracing

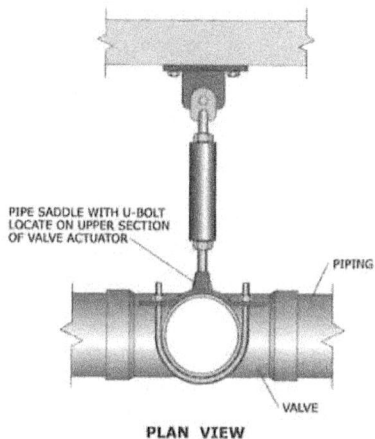

PLAN VIEW

Figure 143: Valve actuator bracing.

Step 1: Run piping as shown on the approved construction documents

 Lay out all attachment points before making final piping connections.

Step 2: Verify lengths used for the pipe bracing

Temporarily fit bracing as necessary.

Step 3: Attach mounting plates to the building structure

For instructions on installing anchors, see Anchors (page 107).

Step 4: Assemble bracing

END OF DETAIL.

ANCHOR SELECTION GUIDE

Power-Actuated	
Description	**Warning**
Threaded Studs Used in cases where the fastened equipment is to be removed later, or where shimming is required. Threaded studs for concrete have a 0.140" to 0.180" shank diameter, with typical penetration of 3/4" (minimum) to 1-1/2" into concrete. Threaded studs for steel plate applications have a 0.140" to 0.180" shank diameter when the steel plate thickness is 3/16" or greater.	**Safety is the primary concern when using power-actuated tools (PAT). PAT tools pose the greatest risk to the operator and others in the area of use. Observe the following safety precautions:** Typically not used for equipment weighing more than 40 pounds. Never allow a tool to be used until the operator is properly trained for the specific tool and application. Never use a tool unless all safety features are functioning properly.
Drive Pins Used to directly fasten equipment for permanent installation. Drive pins used for concrete have a 0.140" to 0.180" shank diameter, with typical penetration of 3/4" minimum to 1-1/2" into concrete. Drive pins for steel plate applications have a 0.140" to 0.180" shank diameter when the steel plate thickness is 3/16" or greater.	Always have the operator and others around wear the proper safety devices. Never use more powerful loads than required for the particular application. Always be aware of the potential of the fastener passing through the substrate or being deflected from its intended target. Make sure that all areas are clear behind and around the target area. Have an action plan in place to properly handle and dispose of misfired loads. Always make sure the tools are low velocity and not standard velocity. (Standard velocity tools are not typically allowed on most jobsites because of the danger.) **Authorities having jurisdiction may require a license to use power-actuated anchors.**

Adhesive	
Description	Warning
Capsule Spin-In Adhesive mixes in the hole when the anchor is drilled by a rotary hammer drill only. Various strengths and types of rods or fasteners can be used. Multiple types of coatings on rods are available. Most commonly used in concrete; some might be suitable for use in other substrates. Most capsules cure quickly compared to epoxy.	Do not over-spin during installation. The rod must have a roof cut end with a single or double 45-degree angle/bevel for mixing. The hole must be clean and dry to achieve the maximum strength. Rod must be clean and must not be disturbed during curing. Many capsules produce strong odors during the curing process. Check the MSDS sheets or with your supervisor to determine if masks are required.
Capsule Hammer-in Adhesive mixes in the hole when the rod is driven by a hammer. Various strengths and types of rods or fasteners can be used. Multiple types of coatings on rods are available. Most commonly used in concrete; some might be suitable for use in other substrates. Most capsules cure quickly compared to epoxy.	The hole must be clean and dry to achieve the maximum strength. Rod must be clean and must not be disturbed during curing. Many capsules produce strong odors during the curing process. Check the MSDS sheets or with your supervisor to determine if masks are required.

Adhesive (cont.)	
Description	Warning
Epoxy Used by mixing two or more components with a mixing nozzle at the point of application. Can be used with multiple forms of fasteners or as an adhesive. Many brands can be used in wet, damp, or dry conditions. Many formulas are allowed for use for USDA food processing areas. Some may be able to be used overhead. Permitted many times in freeze-thaw and severe weather conditions. Allows minimal edge distance and anchor spacing. Typical shelf life greater than that of other adhesives used for anchoring. Not as susceptible to damage from high storage temperatures.	Typically requires long curing times compared to that of other adhesives. Can be virtually odor free or can emit a strong odor, depending on the formula. Can be difficult to apply if the epoxy is thick. Generally not suggested for use at temperatures below 32 degrees F. Most epoxies require holes to be cleaned to obtain maximum values. Check the MSDS sheets or with your supervisor to determine if masks are required.
Acrylic Adhesive Dispenses and cures quickly. Some adhesives can be used overhead. Some adhesives can be installed in damp or water-filled holes. Typically can be used with many fastening devices such as threaded rod, dowels, and anchors.	Many types of acrylics produce a strong odor during the curing process. Others have a minimal odor. Check the MSDS sheets or with your supervisor to determine if masks are required.
Adhesive Undercut Anchors Used in heavy-duty applications where substrate is of poor quality.	Generally purchased from the manufacturer as a complete anchoring system. Any substitution of materials must be authorized before installation.

Externally Threaded	
Description	Warning
Heavy Duty Undercut Used in heavy-duty applications. Typically two types: self-undercutting and adhesive. Self-undercutting types use a special undercutting drill bit similar to heavy-duty sleeve anchors except that they fill a cavity greater than the initial hole diameter.	May require special tools and specific drill bits. Typically cannot be used at variable embedment depths. Can be complicated to install. May be difficult to verify proper installation.
Wedge Anchor The most common concrete anchor for heavy- to light- duty applications. Many configurations are available for most applications. Made from a variety of materials.	Typically designed for static loads and not used with reciprocating engines or in situations where vibrations are present.
Heavy Duty Sleeve Anchor Expansion anchor for heavy-duty requirements.	A large hole is required for this anchor. Some anchors have metric diameters. Some have multiple parts that can be unassembled. If re-assembled improperly, the anchor may not perform properly. If the nut is removed after the stud is inserted in the hole, the stud could be partially separated from the expansion cone, causing a reduction in anchor strength, or be detached from the expansion cone, requiring anchor replacement. These conditions are not visible.

Externally Threaded (cont.)	
Description	Warning
Center Pin Anchor Medium-duty expansion anchor. The anchor is correctly installed when the pin is completely inserted. Installation procedures are simple; no torque is required to set the anchor.	Typically designed for static loads and not used with reciprocating engines, motors or in situations where vibrations are present.
Sleeve Anchor Universal anchor for light- to medium-duty applications. Multiple head designs fit many applications and can be installed in masonry.	A large hole is required for this anchor. Some anchors have metric diameters. Some have multiple parts that can be unassembled. If re-assembled improperly, the anchor may not perform properly. If the nut is removed after the stud is inserted in the hole, the stud could be partially separated from the expansion cone, causing a reduction in anchor strength, or be detached from the expansion cone, requiring anchor replacement. These conditions are not visible.
Concrete Bolts *(continued on next page)* Concrete bolts are recommended for use in dry interior applications. Concrete bolts may be acceptable in some cases for temporary use in exterior applications. Not all concrete bolts used in resisting earthquake or wind loads are appropriate and may be beyond the scope of the anchors abilities.	Maintain equal or greater than minimum edge distance spacing specified for concrete bolts or apply reduction factors if applicable. The embedment depth is the distance from the concrete surface to the bottom of the screw anchor. Maintain minimum slab thickness. Use ONLY a specific manufacturer's recommended method of installation.

Externally Threaded (cont.)

Description	Warning
Concrete Bolts *(cont.)* Concrete bolts should not be subjected to vibratory loads such as those encountered by supports for reciprocating engines, crane loads, and moving loads. Concrete bolts are limited to installation in uncracked concrete masonry. Cracking may occur from anchor location in tension zone of concrete member, and anchors subjected to seismic loads, wind loads, or moving loads. Concrete bolts are limited to non-fire-resistive construction unless appropriate data is submitted demonstrating that anchor performace is maintained in fire-resistive situations.	NEVER substitute one manufacturer's concrete bolt installation instructions for another manufacturer's instructions. DO NOT ASSUME one manufacturer's instructions are the same as another manufacturer's instructions. Some manufacturers use different diameter drill bits for the same nominal diameter concrete bolt. Use the proper diameter drill bit before installation. When a specific screw anchor manufacturer's drill bit diameter is not used, a reduction of capacity must be accounted for in the calculations. Other manufacturers may *recommend* the use of ANSI B212 matched diameter drill bits for use with their concrete bolts. Some manufacturers may allow power tools for installation and others may not. Some screw anchor manufacturers recommend not using an electric impact wrench when re-using the same hole. Some manufacturers allow piloting a new anchor hole; 1 or 2 additional applications or re-uses are possible. If the manufacturer allows the screw anchor to be reused, inspect for excessive wear or the capacity of the screw anchor.

Internally Threaded	
Description	Warning
Internally Threaded Undercut Anchor Used in heavy-duty applications. Typically come in two types: self-undercutting and those using a specialized undercutting drill bit. Anchors have internal threads. Shallow embedment and small edge distances and spacing are possible.	May require special tools and specific drill bits. Typically cannot be used at variable embedment depths. Can be complicated to install. May be difficult to verify proper installation.
Shell Anchor Flush-mount or sub-surface internally threaded anchor for medium- to light-duty applications. Comes in fractional and metric sizes and is available in a variety of materials.	A special setting tool is required and must be supplied by the anchor manufacturer. The setting tool is designed for each anchor size and style.
Others Similar to the wedge concrete anchor and used in heavy- to light-duty applications. Many configurations are available to fit most applications. Made from a variety of materials.	Typically designed for static loads and not used with reciprocating engines or in situations where vibrations are present.

Light Duty Fastenings	
Description	**Warning**
Drive Pin (nail) Anchors (metal and plastic) Light-duty anchor with fast and easy installation in many substrates.	Use only for static loads. Typically not used in overhead applications. **DO NOT USE FOR SEISMIC RESTRAINT**
Concrete Screws Medium- to Light-Duty A variety of lengths and diameters are available. Often used for temporary anchorage.	Typically not used in situations where extensive vibrations are present. Requires the use of a special drill bit (some metric) supplied by the anchor manufacturer.
Special Style Head Wedge (ring) anchor Wedge anchor with integrated connection (head) designed for tie wires or suspended ceilings.	Typically designed for static loads and not used with reciprocating engines or in situations where vibrations are present.
Single and Double Expansion Shields Multi-purpose anchor used in concrete, Concrete Masonry Unit (CMU), brick, or stone. This anchor distributes fairly even pressure, making its use popular in CMU, brick, and natural stone. Typically used in conjunction with machine bolts, which can be removed and replaced.	Anchor material is malleable and the threads can be stripped. **DO NOT USE FOR SEISMIC RESTRAINT**

155

Light Duty Fastenings (cont.)	
Description	Warning
Lead Expansion Anchors Similar to expansion shields, but typically considered light-duty. Many can be used with a variety of screws or bolts. Quick and simple to install. Can be used in concrete, CMU, brick, or stone.	Anchor material is malleable and the threads can be stripped. Anchor should not be used in any applications. **DO NOT USE FOR SEISMIC RESTRAINT**
Toggle or "Molly"-type Anchors Light- to medium-duty anchor with easy installation in many substrates. No drilling is required for some anchor types or in some substrates. Some anchors are supplied with bolts or screws. Anchors are made from variety of materials and colors including plastic, zinc alloys, and steel.	May require a large hole. Anchor may or may not be reusable if the bolt is removed. Severe damage to the substrate can result if these anchors are removed after installation.

GLOSSARY

 Adhesive anchor – A smooth or deformed steel bar or threaded rod, set in a predrilled hole in hardened concrete or masonry (including masonry units and mortar joints) that derives its holding strength from a chemical bonding compound placed between the wall of the hole and the embedded portion of the anchor.

Anchor – A device for connecting equipment and attachments to the building structure.

Attachments – Support systems used to connect equipment, pipe, conduit, or ductwork to the building.

Attachment type – Use of attachments to floors, walls, roofs, ceilings, and vibration isolators.

Bar joist – Ceiling joists supporting intermediate floors or roof made from steel angles and steel bars.

Base plate – A steel plate used for support and anchorage of an angle support or vibration isolator.

Bed joint – A horizontal seam in a brick or concrete block wall. Also see **Head joint**.

Bolt diameter – Thickness or width of the outside of the threaded portion of the bolt.

Building structure – Steel, concrete, masonry and wood members or surfaces that transfer the weight of the building and equipment to the ground.

Bumpers – Angles or other steel shapes with elastomeric padding rigidly mounted to the building structure in a pattern around the equipment base to limit horizontal movement.

Cable brace – A steel cable designed for use as a seismic sway brace for suspended equipment, piping, ductwork, or raceways. Also see **Pre-stretched cable**.

Cant strip – A material used to fill voids in roof flashing.

Cantilevered – A support member connected at one end and unsupported at the other end.

157

Cast-in-place – A steel shape embedded into concrete.

Cast-in-place anchor – A headed steel bolt set within a concrete form before concrete is poured.

Cold joint – An edge between two concrete surfaces.

Construction documents – Drawings, specifications, and manufacturer's instructions (approved by the appropriate design professional) that define the scope of a project and provide detailed information to seismically restrain the equipment, piping, ductwork, or raceways. Also known as blue prints.

Counter flashing – A light-gauge sheet metal folded support or equipment frame to shed water or snow onto the roof.

Curb – Raised or enclosed framework that supports equipment.

Cure – To gain internal strength over time to withstand external forces.

Cure time – The total time it takes for the material to be at an absolute full load capacity.

D

Design professional - The responsible party, recognized by the authority having jurisdiction, working within their scope of qualifications and providing design services.

Differential movement – The movement between two objects or surfaces.

E

Edge distance – The distance between a concrete anchor and the edge of a concrete surface or concrete cold joint.

Elastomeric – A material with flexibility in all directions that will return to its original shape if removed from its environment.

Elastomeric pad - A resilient natural or synthetic rubber like pad used to reduce sound, shock and high frequency vibrations.

Embedded plate - A steel plate set into concrete to permit a welded attachment.

Embedment – How far a post-installed anchor is inserted into a hole in concrete or wood after the anchor is set in place and torqued.

Embedment depth – See **Embedment**.

Enclosure – A case or housing to protect electrical components.

Equipment – Any mechanical or electrical component.

Expansion anchor – A post-installed anchor that uses some form of wedge or shell held against the edge of a drilled hole with friction.

F

Ferrule – A small metal tube that can be crimped around steel cables.

Fiberglass pad - A pad with a core of resilient fiberglass material covered by a moisture resistent resilient shell used to reduce sound, shock and high frequency vibrations.

Fillet weld – A weld between two pieces of steel where the welded surfaces are at right angles.

Flashing – Metal, asphalt, or elastomeric material with one or more layers surrounding a roof penetration specifically designed to weatherproof the building.

Flexible connector – A connector designed to allow slight movement between a piece of equipment, component, or system and another system in the event of an earthquake.

Flexible mounted equipment – A piece of equipment supported on or from a vibration isolator.

G

Gel time – A specified amount of time for an adhesive to form a jelly-like substance with strength to hold its own weight or the weight of a light steel anchor.

Grommet – A rubber or elastomeric bushing-shaped ring that may be used in restrained springs, snubbers, or with bolts to provide a cushioned or flexible connection.

Groove joint – A mechanical connection between two pipe sections using a tongue-and-groove configuration and elastomeric gasket.

Hand tight – The force applied by hand or with hand tools to bring two or more materials firmly in contact without a space.

Head joint – A vertical joint between two concrete blocks in a block wall or two bricks in a brick wall. Also see **Bed joint**.

Headed stud – A large bolt with a threaded shaft and a hexagonal-shaped bolt head typically used for embedment into concrete surfaces or in-filled concrete walls.

Height-saving bracket – A bracket used to accommodate the height of spring isolators without raising the equipment base more than a few inches.

Housed spring – A spring isolator with steel guides usually separated by an elastomeric sheet located on two opposite sides of the spring.

Housed spring isolator - A steel coil spring designed to be loaded in compression along the axis of the spring. Housed springs have a two-piece housing and mounting stud for the purpose of leveling equipment. The housing has limited lateral restraining capabilities and cannot resist uplift.

Housekeeping pad – A concrete pad under equipment that raises the elevation of the equipment above the building structure or structural slab.

 Inertia base – A heavily weighted base, usually made of concrete, that weighs more than the equipment it supports.

In-filled block – A concrete block wall whose cells are reinforced with rebar and filled with a sand-grout mixture.

Inlet – The location or connection to equipment where a substance such as water or air enters the equipment.

In-line equipment - Equipment connected to ducts or pipes allowing the continuous flow of the medium inside the ducts or pipes.

Isolators – See **Vibration isolators**.

L

Leveling stanchions – See **Stanchions.**

Load path – Seismic support of equipment and internal components that can be traced through connections and support steel to the building structure.

Load transfer angles – Angles bolted to equipment and to the building structure, transferring the weight and earthquake load through the angles to the building structure.

Longitudinal brace – A brace that restrains pipes, ducts, or raceways parallel to the longitudinal direction of the pipe, duct run, or raceway.

M

Moment-resistant - The ability of an object to resist a force that would cause it's rotation or movement around a point or axis.

N

No-hub pipe – Pipe designed for connections that do not interlock or permanently join.

Nominal diameter – The diameter across the outer-most edges of a bolt or threaded rod.

O

Open spring – A spring isolator with a bolt attachment at the top of the spring for connecting to equipment without any horizontal support.

Open spring isolator - A steel coil spring designed to be loaded in compression along the axis of the spring. Open springs have a plate or cup on the bottom and top with a mounting stud.

Outlet – The location or connection to equipment where a substance such as water or air exits the equipment.

Oversized hole - Bolt holes greater than standard bolt holes allowed by industry standards.

P

Pitch pocket - The frame on a roof contouring the sealing material protecting the penetrations or surface-mounted equipment supports on an unprotected building structure.

Plenum – An enclosed space usually made from galvanized sheet steel allowing airflow from one duct system to another; the entrance to and/or exit from a fan or air handling unit.

Plug weld – The weld of a plate or base plate to another metal surface where a plate is perforated with one or more holes, which are then filled with the weld filler material.

Point load – Weight and seismic forces that are focused to a single point connection to the building structure.

Post and beam – An elevated structure usually made from beams resting on posts or stanchions connected to the building structure.

Post-installed anchor – Anchors installed after the concrete has reached its specified strength.

Post-tension (pre-stressed) building – A concrete building structure surface with internal steel cables that are stretched and restrained to permanently compress the concrete surface.

Pre-stressed beam - A concrete beam bonded to steel in tensionin the form of a beam.

Pre-stretched cable – Cable that is stretched after it is manufactured.

 Rated spring deflection – The dimension a spring will compress when the weight of equipment is applied.

Rehabilitation – A new installation within an existing facility.

Restrained elastomeric mount - An elastomeric mount with an integrated seismic restraining mechanism that limits movement in all horizontal and vertical directions

Restrained spring – A vibration isolator containing a spring enclosed in a welded or bolted steel housing that limits the movement of the spring equipment attachment in all directions.

Restrained spring isolator - A steel coil spring designed to be loaded in compression along the axis of the spring. Restrained springs have a housing with integrated seismic restraining mechanisms, which limit movement in all horizontal and vertical directions.

Rigid-mounted equipment – Equipment solidly braced or bolted directly to the building structure without vibration isolation.

S

Screen – A tube of steel wire mesh used as an adhesive anchor for anchoring to block or brick walls.

Seismic cable – A steel or stainless steel braided rope.

Seismic restraint device – An attachment device designed to restrict movement of equipment during an earthquake.

Seismic restraint device submittals – Documents created by contractors or vendors describing the means and methods for installing seismic restraint devices and submitted for design approval.

Seismic rod clamp – A clamping device for attaching rod stiffeners to a vertical threaded rod.

Seismic separation joint - A space provided between buildings or portions of a building to prevent contact caused by differential movement during an earthquake. For piping, ductwork and conduits crossing the seismic joint, the systems are connected with a flexible component that allows for differential movement at least twice the width of the seismic separation joint.

Self-drilling – A special type of concrete shell anchor with cutting teeth for drilling into concrete.

Self-tapping – Either a sheet metal screw with blades on the end (similar to a drill bit), allowing the screw to drill a hole and embed itself into a steel shape, or a concrete screw with a point and specially designed threads allowing the screw to grip the concrete and embed itself into the concrete.

Set time – The specific time required for material to harden when a light load may be applied.

Shallow concrete anchor – Any anchor with an embedment depth measuring less than 1/8th of its diameter.

Sheet steel housings – Sheet steel that fully or partially encloses a piece of equipment.

Shim – A thin wedge of material used to fill a space.

Snubber – A seismic restraint device used on isolated systems with an air gap and elastomeric bushing or oil-filled hydraulic cylinder (shock absorber) restricting the rapid motion of a pipe.

Snug tight – The force applied by hand or hand tools to bring two or more materials firmly in contact without a space.

Solid brace – A steel angle or strut channel designed for use as a seismic sway brace for suspended equipment, piping, ductwork, or raceways.

Spring-isolated – See **Vibration-isolated.**

Stanchions – Columns or short structural steel shapes placed vertically that connect to equipment bases or horizontal structural steel frames to provide equipment support.

Structural steel shapes – A manufactured steel component in a variety of shapes.

Strut – A manufactured steel shape in various U-shaped patterns and sizes.

Strut frame – Steel framing made from strut members that act as a support to transfer the equipment weight to the building structure. See **Strut**.

Sway brace – Solid braces or cable braces that provide seismic restraint.

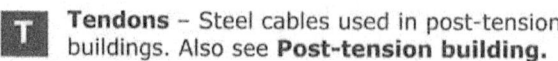

Tendons – Steel cables used in post-tension buildings. Also see **Post-tension building.**

Thimble – A metal spacer used on a cable to protect it from being bent and damaged.

Transverse brace – A brace that restrains pipes, ducts, or raceways perpendicular to the longitudinal direction.

Toe of the weld – The edge of a fillet weld.

Torque – A turning force around a bolt applied by twisting a bolt head or nut so the components will not separate.

Turn-of-nut method – A process to properly torque a bolt without a special tool like a calibrated torque wrench. See **Hand tight** and **Snug tight**.

VAV boxes – A terminal unit or plenum with an internal damper and control actuator that can vary airflow quantities.

Vibration-isolated – Used to describe a system that is separated from the building structure with devices designed to reduce vibration and noise transmission into the building structure.

Vibration isolators – Components containing resilient elements such as steel springs, air springs, molded pre-compressed fiberglass or elastomeric pads used to separate vibrating equipment, piping and ductwork from the building structure.

Web – A thin metal strip in a structural steel shape.

Weld base material – The material composition of an item being welded.

WPS - Weld Procedure Specification is required for all welding in accordance with American Welding Society D1.1. The WPS defines the essential variables and their limits for the weld and must be in the vicinity where the weld is occuring.

INDEX

A

flashing 51, 158
flexible connectors 140-141
floor-mounted
ducts 8-9, 41-43
piping 12-13, 70-72

G

gap, *see acceptable gap*
grouted plate 113

H

heat-exchanger 14, 100-101
hole
cleaning 111
clearance 116
depth 110
drilling 109-111
humidifier 10

I

in-line
air separator 99-100
duct-mounted equipment 10-11
fan 11
heat exchanger 100-101
pipe-mounted equipment 14-15
pump 97-98
interior penetrations
duct 8-9, 52-53
pipe 12-13, 82-83

L

lag bolts 107, 115-116, 117, 123
longitudinal bracing 6-7, 17, 23, 27, 29, 34, 61, 62, 160

M

masonry and drywall anchors *see anchors*
motorized damper 10

V

W

FEMA

FEMA P-414
Catalog No. 10148-1

www.ingramcontent.com/pod-product-compliance
Lightning Source LLC
Chambersburg PA
CBHW051505170526
45166CB00001B/393